On the cover. 22 March 2007, a U.S. Special Forces soldier instructs Malian troops in counterterrorism tactics through a translator. Originally published on Vanityfair.com, 10 September 2007. Photograph by Justin Bishop.

JOINT SPECIAL OPERATIONS
JSOU
UNIVERSITY

Africa: Irregular Warfare on the Dark Continent

John B. Alexander

JSOU Report 09-5
The JSOU Press
Hurlburt Field, Florida
2009

Comments about this publication are invited and should be forwarded to Director, Strategic Studies Department, Joint Special Operations University, 357 Tully Street, Alison Building, Hurlburt Field, Florida 32544. Copies of this publication may be obtained by calling JSOU at 850-884-1569; FAX 850-884-3917.

****

The JSOU Strategic Studies Department is currently accepting written works relevant to special operations for potential publication. For more information please contact Mr. Jim Anderson, JSOU Director of Research, at 850-884-1569, DSN 579-1569, james.d.anderson@hurlburt.af.mil. Thank you for your interest in the JSOU Press.

****

ISBN 1-933749-34-2

Recent Publications of the JSOU Press

Educating for Strategic Thinking in the SOF Community, January 2007, Harry R. Yarger

The Israeli Approach to Irregular Warfare and Implications for the U.S., February 2007, Thomas H. Henriksen

Psychological Operations: Learning Is Not a Defense Science Project, March 2007, Curtis D. Boyd

2007 JSOU and NDIA SO/LIC Division Essays, April 2007

Hunting Leadership Targets in Counterinsurgency and Counterterrorist Operations, June 2007, Graham H. Turbiville, Jr.

Executive Report, JSOU Second Annual Symposium (30 April–3 May 2007)

A Theory of Special Operations, October 2007, Robert G. Spulak, Jr.

Block by Block: Civic Action in the Battle of Baghdad, November 2007, Adrian T. Bogart III

Private Security Infrastructure Abroad, November 2007, Graham H. Turbiville, Jr.

Intelligence in Denied Areas, December 2007, Russell D. Howard

Is Leaving the Middle East a Viable Option, January 2008, Thomas H. Henriksen

Retaining a Precarious Value as Special Operations Go Mainstream, February 2008, Jessica Glicken Turnley

Disrupting Threat Finances, April 2008, Wesley J.L. Anderson

USSOCOM Research Topics 2009

India's Northeast: The Frontier in Ferment, September 2008, Prakash Singh

What Really Happened in Nortern Ireland's Counterinsurgency, October 2008, Thomas H. Henriksen

Guerrilla Counterintelligence: Insurgent Approaches to Neutralizing Adversary Intelligence Operations, January 2009, Graham H. Turbiville, Jr.

Policing and Law Enforcement in COIN—the Thick Blue Line, February 2009, Joseph D. Celeski

Contemporary Security Challenges: Irregular Warfare and Indirect Approaches, February 2009, Richard D. Newton, Travis L. Homiak, Kelly H. Smith, Isaac J. Peltier, and D. Jonathan White

Special Operations Forces Interagency Counterterrorism Reference Manual, March 2009

The Arabian Gulf and Security Policy: The Past as Present, the Present as Future, April 2009, Roby C. Barrett

Foreword

Dr. John Alexander's current JSOU Press monograph provides an assessment of the African continent with a particular focus on how Special Operations Forces (SOF) may need to operate and how the local environment impacts these operations. He provides a strategic overview and assessment of current conditions on the continent, identifies key concerns and issues, and discusses key players. For a variety of reasons, Africa is, and will remain, a region of critical importance not only to the United States but also to other regions and countries of the world.

This monograph will be very useful for SOF planners and operators as well as conventional forces and interagency partners working in the region. Dr. Alexander provides insight into major issues concerning Africa from tribalism, resource competition, and political instability to terrorist and criminal network linkages. The author bases his assessment on a review of historical trends as well as the impact of current political and geostrategic events. His realistic assessment indicates the region is likely to remain volatile for the short-to-medium timeframe.

Conflicts in the Sudan, Congo, and the Horn of Africa combined with instability or potential instability within other key African countries will require external support and assistance from the United States and its allies to work out a viable solution to their problems. Much of this assistance will not fall into the military realm. Unfortunately much of the instability will turn into conflict and may require multinational military intervention. This tension between assistance and intervention is the underlying conundrum facing strategists and planners—Africa is too important to the world to continue to ignore its significant challenges and conflicts.

Michael C. McMahon, Lt Col USAF
Director, JSOU Strategic Studies Department

About the Author

Dr. John Alexander has been a leading advocate for the development of nonlethal weapons since he created renewed interest in the field in 1989. In 2003 he served as a mentor to Afghan Ministry of Defense senior officials through the Office of Military Cooperation Afghanistan, (Coalition Forces) Kabul. He now serves as a senior fellow with the JSOU Strategic Studies Department.

Entering the U.S. Army as a private in 1956, he rose through the ranks to sergeant first class, attended Officer Candidate School (OCS), and was an infantry colonel in 1988 when he retired. During his varied career, Dr. Alexander held many key positions in special operations, intelligence, and research and development. From 1966 through early 1969 he commanded Special Forces "A" teams in Vietnam and Thailand. His last military assignment was as director, Advanced System Concepts Office, U.S. Army Laboratory Command. After retiring from the Army, Dr. Alexander joined Los Alamos National Laboratory where he was instrumental in developing the concept of nonlethal defense. As a program manager, he conducted nonlethal warfare briefings at the highest levels of government including the White House staff, National Security Council, members of Congress, director of Central Intelligence, and senior defense officials. He has met with heads of industry and presented at academic institutions, including Columbia, Harvard, and the Massachusetts Institute of Technology (MIT). Dr. Alexander has spoken on all continents, including to the German Bundestag and members of the Chemical Weapons Convention in The Hague.

Dr. Alexander organized and chaired six major conferences on nonlethal warfare and served as a U.S. delegate to four NATO studies on the topic. As a member of the first Council on Foreign Relations nonlethal warfare study, he was instrumental in influencing the report that is credited with causing the DoD to create a formal nonlethal weapons policy in July 1996. He was a distinguished guest lecturer at the U.S. Air Force Air University. He has advised the Central Intelligence Agency, USSOCOM, and the National Intelligence Council.

Dr. Alexander wrote the seminal material on nonlethal warfare. He has published articles in *Harvard International Review*, *Jane's International Defense Review*, *The Boston Globe*, *The Futurist*, *The Washington Post*, and several other journals. Articles about him and his work can be found in *The Wall Street Journal*, *The New York Times*, *Newsweek*, *The Sunday Times* (London), *Panorama* (Italy), *Konrad* (Germany), *The LA Times*, *Wired Magazine*, *GQ*, *Scientific American*, and the *Bulletin of Atomic Scientists*. He has appeared on U.S. television including *Dateline*, *Fox News*, *Larry King Live*, *CNN*, *MSNBC*, *Newsweek*, and on international television in Australia, Denmark, France, Germany, Italy, Brazil, Switzerland, the Netherlands, United Kingdom, and Japan. His April 2006 JSOU publication was *The Changing Nature of Warfare, the Factors Mediating Future Conflict, and Implications for SOF.*

Dr. Alexander received a B.GS. in Sociology from the University of Nebraska at Omaha, an M.A. in Education from Pepperdine University, and a Ph.D. in Education from Walden University. He later attended the Anderson School of Management at University of California, Los Angeles (UCLA), the Sloan School of Management at MIT, and the Kennedy School of Government general officer program "National and International Security for Senior Executives" at Harvard University. In addition to many military awards for valor and service, *Aviation Week & Space Technology* selected him as a 1993 Aerospace Laureate and in 1997 inducted him into the Hall of Fame at the National Air and Space Museum in Washington. He received a Department of Energy Award of Excellence for the Nuclear Weapons Program in 1994 and is listed in *Who's Who in America*, *Who's Who in Science and Engineering*, and *American Men and Women of Science*. In 2001 he was named to the OCS Hall of Fame at Fort Benning, Georgia. During 2001 and 2002 he was a member of the National Research Council Committee for Assessment of Nonlethal Weapons Science and Technology.

Currently Dr. Alexander serves with the Army Science Board and is a private consultant. His books include *The Warrior's Edge* (William Morrow & Co, 1990); *Future War* with foreword by Tom Clancy (St. Martin's Press, 1999); and the sequel *Winning the War* (2003).

Africa: Irregular Warfare on the Dark Continent

Irregular Warfare: Introduction

If there ever was an environment that will test irregular warfare (IW) concepts, it is Africa. There are protracted conflicts—some that come with centuries of history, intricate tribal animosities fueled by greed and corruption and exacerbated by outside powers. While there are countries with large armies, none boast the range of high technology found in other areas of the world. Most of the casualties—and the numbers are horrific—come from small arms, even from spears and machetes. Present in some areas on the continent are some artillery pieces and fighter aircraft, but nothing that would counter the U.S. military on a force-on-force confrontation. Yet our experience shows that even overwhelming technology may not win the day. Given that Africa is rising rapidly in our national interests, we must explore its complex environment.[1]

Enter Africa Command

"We didn't just discover Africa," a senior Joint commander told a small group assembled for the regional meeting following the International Special Operations Forces conference. The intent seemed to be to assure those present that indeed, Africa had been known to special operations for a long time.[2]

While it is true that special operations forces (SOF) have conducted operations in many African countries for decades, they have suffered from a lack of a coherent United States policy regarding that continent. Even with a new command dedicated to focus attention in the area, a senior Defense Department official described Africa as a "minor sideshow" in global counterterrorism operations. He also bluntly noted that people like him, who had very little understanding of the issues, would make decisions about U.S. policy in the region. Speaking to a large audience for a conference at the University of Southern California with many American and African ambassadors present, the DoD official stated that Africa Command was "an experiment in government organization," and the military objective was to "keep U.S. combat troops off the continent for at least fifty years." [3]

At the same University of Southern California conference, an ambassador assigned to AFRICOM stated that the existence of the new command was "an acknowledgement for the growing strategic importance of Africa." [4] In the dinner speech it was also noted that American emphasis would be larger than humanitarian assistance, which frequently has been required for decades of both natural and human generated disasters. The American objective would be to support the African Union as it continues to develop. The oft-used buzzwords infer that we need to help African nations *build capacity*—a concept never really defined.

Africa, map courtesy Wikipedia Commons.

If this discussion seems vague and convoluted, it is—just as the development of AFRICOM is vague and convoluted. Continentally homeless, the AFRICOM headquarters is located in Stuttgart, Germany, also home to European Command (EUCOM). Although several African nations were approached about establishing an AFRICOM headquarters, all refused. In reality, there is great trepidation about why the command was formed and what U.S. intentions are in the region. Due to some poorly constructed initial information releases about formation of this command, substantial effort has been required in attempting to make nations in Africa, as well as others around the world, comfortable with that decision. This is far from accomplished and will remain a priority mission for some time to come.

Special Operations Considerations

This paper is not about the trials and tribulations of Africa Command. It is designed to provide the reader with general information about the continent, specifically as it relates to irregular warfare and the challenges for SOF. In fact, operations in Africa will require the full spectrum of SOF missions from direct action against Al Qaeda terrorists to internal development for several countries. History has shown that humanitarian assistance is likely to be required and often on massive scales. In fact SOF elements have established precedence in these operations.[5] While this is not a specified core mission of USSOCOM, it is one that that will be assigned for a variety of reasons, including prior geographic orientation, experience in the area, ability to respond quickly and adapt to exigent circumstances, as well as the capabilities resident in the active duty Civil Affairs units.

Worth considering are the factors employed by the U.S. Joint Forces Command when assessing the Joint Operational Environment (JOE). The eight factors include demographics, globalization, economics, energy, climate change and natural disasters, pandemics, and cyber and space.[6] Africa is unique, fragile, and vulnerable to many extraneous influences that can have deep impact on the future for residents and the rest of the world alike. Thus an attempt has been made to cover each of the areas referenced in JOE 2008.

On the continent there are significant armed groups that threaten regional stability in several sectors. From the Sahel—the semi-arid transition zone between the Sahara to its north and the less arid, more tropical lands to its south—there are terrorist organizations that can and do project power well beyond the African shores and have been instrumental in attacks in Europe. Internally there are long running conflicts, some of genocidal proportions, and poverty that clearly exceed the capabilities of the African Union to resolve.

In another monograph addressing emerging events in South and Central America, I opened with a quote by a U.S. Army lieutenant colonel assigned to the Deputy Chief of Staff for Operations (G3). When apprized of the rather dire circumstance arising south of our borders, he simply stated, "Why should I care?"[7] For most Americans, including many military officers, that same comment would hold true for Africa. With the U.S. Department of Defense terminally focused on the Middle East, most of Africa is off their radar screens. They are far from alone and most Americans have only the vaguest notion of what Africa is like. Their simplistic opinions are often based on trite movies or television programs that frequently highlight the spectacular scenery and vicious interaction with dangerous wildlife. While travelogues are worthwhile, rarely do they convey a complete, or even adequate, description of the complex situations that constitute African realities.

... most Americans have only the vaguest notion of what Africa is like.

The original reference to Africa as *The Dark Continent* was associated with colonial popularization of the territories during their early exploration. In many ways the connotation was perceived quite negatively. Today, however, there is a more literal application of the term. In sub-Saharan Africa, not counting South Africa, 75 percent of the households do not have access to an electrical grid. That equates to 550 million people who at night, live in the dark. The implications for development are staggering and energy poverty is becoming a key factor in participation in the global economy.[8]

There are many compelling factors that complicate both life in, and geopolitical interactions with, most African nations. These include extreme poverty, rampant diseases (including an AIDS epidemic), massive political corruption, tribalism with historical animosities, inhospitable geographic environments, illogical national borders, burgeoning population growth, inadequate education systems, and centuries of foreign interventions. Also, despite the oft quoted "never again" that followed revelation of the Nazi death camps, Africans continue to live in a world tolerant of repeated genocidal level of casualties.

The Continent: History and Complexities

Despite a multitude of drawbacks, the reality is that Africa is rapidly gaining in strategic significance and has attracted new, or renewed, interest in many sectors. In addition to the Europeans, who have traditionally viewed Africa as their domain, both India and China are investing heavily in many countries, with an eye toward the vast mineral resources that can be found there. As emerging industrial powers, the future needs for energy and raw materials are concerns for which they are planning.

Africa is a huge continent marked by unparalleled diversity and a complex human history that literally emanates from the origin of life of our species. Paleontological evidence shows that humans likely originated over two million years ago in the Olduvai Gorge region of the Great Rift Valley, in what is now known as Tanzania. Tests on mitochondrial DNA suggest that most humans throughout the world today can trace their ancestry to seven females from that area. Interestingly, it appears that those early hominids were fiercely territorial and engaged in savage intertribal combat for survival. It would appear that not much has changed in the intervening twenty-plus millennia.

There are many factors that combine to make Africa the enigmatic milieu that it has become today. While current American efforts note a desire to move forward, Africa's Byzantine-like history cannot be ignored. Presently, over 400 tribes comprise the native population base. Centuries of invasions and external foreign domination determined the boundaries of most countries and contributed to introduction of other social groups. Those geographic delineations rarely took into account the ethnic or tribal makeup of the local people. As a result, social tensions abound and are constantly exacerbated by current political decisions made for the benefit of a dominant group at the expense of others. Violent outbursts continue today. Within the past few decades such violence has led to the death or displacement of many millions of people, and the conflict is far from over.

Christopher Clapham is the editor of the *Journal of Modern African Studies* at Cambridge University. Reporting for the Royal United Services Institute in 2007, he stated, "Africa is an extremely difficult continent to govern, simply because of its huge distances, generally dispersed population, weak infrastructure, and extremely uneven distribution of resources. All of Africa's rulers, whether pre-colonial, colonial, or post-colonial have faced essentially the same problems in exerting effective control over this large and poor landmass and its often mobile people." He then notes correctly that the "extremely problematic nature of many of the states artificially created by colonialism and bequeathed to the continents new rulers at independence." [9]

Tribes and Tribalism

Before entering this discussion of an important issue, a comment about the concept of tribes is prudent. The term *tribe* generally connotes groupings of people who have certain common characteristics or unifying factors. These may include language, culture, genetics, religion, and sometimes regional association. In fact, there is no agreed-upon definition of what constitutes a tribe as there are always exceptions to the parameters mentioned. For example, some groups have multiple languages and may not share a specified region. There is an element of academic social engineers that has attempted to make the use of the term *tribe* a racial emotive that promotes stereotyping, and is thus politically incorrect.[10] Further discussion follows.[11] Despite semantic controversy, and accepting that *tribe* is an imperfect word, the fundamental concept is useful, and possibly essential, for describing many of the social interactions that occur in Africa today.

In general, the concept of tribalism is not well understood by Americans. Arguably, the United States has become the most homogenized society in the world, though that construct is experiencing considerable stress. Former and existing racial tensions are demarcated along large groupings such as black or African-Americans, Hispanic, Native American, and whites. Rarely, is further subdivision annotated. Most of the African slaves shipped to the New World landed in South America. However, a substantial number did come to North America and became the basis of our current African-American population of about 40 million people. Today, few people who call themselves African-Americans can account for their original tribal identity. Once the slaves arrived, families were split apart and little attempt was made to track former identities. Over time, after emancipation, and especially since the success of the civil rights movement, most African-Americans have assimilated into the more generic common society. While vestiges of racism are acknowledged, American society as a whole has greatly evolved into a technologically developed nation in which advancement in all sectors can be sought.

That is not true for many areas in Africa today. For a very large segment of the population currently living on the continent, tribal identity is still of primary importance. That designation often delineates who you are, with whom you can associate (especially in marriage), what jobs you can hold, what education is available to you, and under extreme circumstances, whether you live or die.

Rwanda. A recent arrant example was the genocidal civil war that occurred in Rwanda in 1994. In 100 days of carnage, the minority Tutsis were systematically hunted down and exterminated by members of the Hutu majority. From April through mid-July hundreds of thousands of Tutsis were systematically shot, clubbed, or hacked to death, with many others burned alive. The number killed will never be known. Conservative estimates place the body count at 800,000, with other analysts taking the toll to over a million fatalities.[12] Despite the fact that the tribes were very similar, in that they spoke the same language and followed the same customs, great animosity developed. That was primarily because of foreign intervention—in this case the Belgians who earlier had bestowed favoritism on the Tutsis. Of specific concern to observers should be the vicious nature of these killings and that the violence seemingly erupted almost instantaneously. Death came equally to men, women, and children, often perpetrated with intentional, excessive pain. Though the provocations had been festering for decades it was single event, the downing of President Juvenal Habyarimana's airplane, that triggered the slaughter.

While 1994 Rwanda stands as a highly visible exemple of tribalism at its worst, it is far from a singularity. In fact, based on personal observations in that country in 2007, it is predictable that the civil war could reignite at any time with little provocation.[13] Other observers also have noted that these issues are far from totally resolved and that the *bapfuye buhagazi* (walking dead) will never manage to forget the atrocities that took place here.[14]

Kenya. A more recent example of intertribal violence precipitated after the contested elections held in Kenya on 27 December 2007. Immediately following the election, the incumbent, President Mwai Kibaki, was declared the winner. Officials he had previously appointed announced the results, which were immediately rejected by many people. International observers supported the opposition party's allegation that the vote had been manipulated. Their candidate, Raila Odinga, complained about fraudulent vote counting and violence quickly ensued. The conflict was divided along tribal lines, which in turn mirrored the support for the respective political parties.[15] In this case it was Kikuyu people, primarily from the Rift Valley Province and the Nairobi slums who were targeted. They had supported Kibaki, who is also a Kikuyu. These spontaneous riots led to between 800 and 1500 fatalities and displacement of over a half million people, many of whom had their homes destroyed.[16] A number of prominent people, including members of parliament, were killed, but the worst single incident was when rioters set fire to a church and burned 35 Kikuyus alive.[17]

Nairobi, Kenya long has been noted for violent crime. The U.S. State Department has issued warnings to Americans traveling there about violent crime in that city.[18] At one point they recommended that visitors carry several hundred dollars so that, if carjacked, they would have money to give the robbers. The warnings noted that the consequence for incomplete immediate compliance, or lack of money on demand, likely was death. The crime rate was so high that the city acquired the title Nia*rob*ia. However, that crime was driven by poverty and was an equal-opportunity victimization regardless of race or tribe. The recent politically motivated unrest is worth noting for two reasons: (1) it was tribally delineated and (2) the speed with which relative stability deteriorated.

At the heart of Kenya's crime is the Mungiki sect which now claims as many as a million members across the nation.[19] The Mungiki, meaning "multitude" in Kikuyu, are believed to be a derivative of the fearsome Mau Mau that fought against British colonial government. Swearing an oath of secrecy—betrayal punishable by death—the gang members are comprised mainly of youths. They began as a hardline offshoot of the Tent of the Living God, which rejected Western values and wanted to return to tribal rituals. The Mungiki are a criminal element that is involved in murder, kidnappings, extortion, and levying protection fees against the poor in the slums where they dwell.[20]

In some slum areas of Nairobi, especially Mathare in the eastern part of the city, Mungiki followers "reign supreme," and their influence is increasing in the central and Rift Valley provinces.[21] Their tactics include extremely gruesome murders, sometimes leaving heads of victims placed on pikes, a la the Mau Mau before them. One such murder was the beheading of the son of Kimwatu Kinyungu, a close friend of President Kibaki. As a statement of power, the head was dumped outside the president's home. As a result, there has been a major crack down on the Mungiki. Between June 2007 and December 2008, more than 600 alleged members have been killed. Many others have been arrested and simply disappeared, which has raised concerns among some local civil libertarians. While police officials deny charges, they have been quoted as saying "the funerals of Mungiki youth would become a common occurrence." [22]

Zimbabwe. Another country generating tribal-related problems is Zimbabwe. Formally known as Rhodesia, there have been long standing tensions between the ruling party, which is predominantly Shona, and the southern minority Ndeble, who are ethnically related to the Zulus of South Africa. Morgan Tsvangirai, the opposition leader, was born in Guto in South Rhodesia.[23] On 29 March 2008 presidential elections were held and it was widely believed that Tsvangirai had rightfully unseated the long-time incumbent, Robert Mugabe, who has held power since 1980. Emerging as a leader, Mugabe was as a guerilla fighter in the rebellion against the white rule in Rhodesia. After the rebellion, he and his Zimbabwe African National Union (ZANU) party, assumed power. Later, in 2002, following 2 years of periodic violence, Mugabe expropriated most of the white-owned farmland on relatively short notice.[24] That caused nearly 3,000 of the agriculturally experienced white residents to leave the country, turned international opinion against him, and precipitated a downward economic spiral. His policies led to severe food and oil shortages with pre-election inflation reaching as high as 9,000 percent and possibly more. In turn, this caused a larger number of citizens of Zimbabwe to flee the country, many moving to the relative stability of South Africa. Adding to the country's woes and instability was a cholera outbreak. Beginning in August 2008, this dreaded disease spread across the land, exacerbated

Robert Mugabe attends the 12th African Union Summit 2 February 2009 in Addis Ababa, Ethiopia. U.S. Navy photo by Mass Communication Specialist 2nd Class Jesse B. Awalt/Released.

by the government's inadequate policies. While normally cholera can be rapidly contained, through the end of March 2009 more than 93,000 cases had been reported including over 4000 fatalities.[25]

The elections did not determine which party would rule. Mugabe refused to accept the results, but did offer accommodations, such as a power-sharing arrangement. The opposition considered the offers. However, Mugabe's concept of power sharing meant that ZANU retained control of all of the key influential elements of governance. At the time of this writing, the final determination of the situation is unclear. Mugabe, supported by his ZANU party, holds a significant advantage when it comes to use of force. He commands the loyalty of the armed forces and has been successful in gaining assistance from China in buying new weapons. A $200 million package for fighter aircraft and military vehicles was concluded despite an arms embargo against Zimbabwe.[26]

The internal situation has continued to deteriorate. In June 2008 Zimbabwe's unemployment rate hit 80 percent while inflation was an astronomical 11,250,000 percent.[27] That level of economic instability is unsustainable and dire consequences, such as armed rebellion or external intervention, are predictable.

But the impact of Mugabe's incompetent leadership did not stop at the country's borders. As instability increased in Zimbabwe, pressure also mounted in South Africa, Mozambique, and other neighboring countries. Though Southern Africa recently has been relatively quiet, riots broke out in many of the major cities. The violence was focused against the foreigners who were settling in their midst. The murders and other attacks were cited as examples of xenophobia. Most of those foreigners were from Zimbabwe, thus of different tribes than the residents of Johannesburg, West Cape, Kwazulu-Natal, Limpopo and other cities affected. Of course, another component of this tense situation was competition for jobs, and outsiders are always considered suspect under such situations. The bright spot was that after dozens of murders, and displacement of tens of thousands foreigners, some South African residents did step up and protest against the violence.[28]

For much of Africa it is traditional tribal or clan laws, customs, and practices that have created and maintained some measure of manageable conflict. However, these strictures were altered irrevocably by colonial "civilizing" measures and detribalization initiatives which were instituted solely for the best interests of the occupying powers. Then there was further social disruption caused by the withdrawal of colonial regimes and their post-colonial involvement that still continues. The complexity is more than just traditional tribalism, which can be disruptive enough. Often a mutated tribalism has to be examined on a case-by-case basis.[29]

Tribalism is not exclusively African. However, it is prevalent throughout a majority of the continent. The implications run far deeper than most Americans can possibly imagine. It is but one factor that must be considered by SOF elements operating in Africa, albeit a very complex one.

Natural Resources

The continent of Africa has abundant natural resources. However, those resources have proven to be both a blessing and a curse. Minerals that can be exported have historically brought foreign invaders and colonialists. In the post-colonial period, as African governments gained independence and became somewhat more sophisticated, they often exploited those resources at the expense of the local population. In some cases, even the presiding governments joined in the deceit with public wealth being transferred into the private bank accounts of the country's leaders. Kleptocracy and other corruption issues are related, are discussed in the section on corruption (page 47), and arise several times later in this paper, as they are quite a significant problem for SOF operations.

Among the best known of the exported natural resources is diamonds. While diamonds are not rare and are found on almost every continent, the discovery in South Africa in the 1870s of what became known as the Kimberly mines transformed the industry. As most of Africa was then under colonial rule, diamonds became a point of contention. Most importantly, the wealth from diamonds predominantly went to the white mine owners. Soon, diamonds were found in southwestern, central and western African countries. Today there are mines in South Africa, Lesotho, Botswana, Namibia, the Democratic Republic of Congo, Sierra Leone, Angola, and Liberia.

While steps have been taken to bring the authenticity of natural diamonds under control, there are still concerns about the illegal market for blood diamonds, also known as conflict diamonds. The mining of these stones in Sierra Leone, and other countries along the west coast of central Africa, had tragic consequences leading to open warfare. The results can still be seen today with over 20,000 people who had their arms chopped off with machetes wielded by guerillas of the Revolutionary United Front.[30]

Awareness of the current status of the diamond industry, and especially the availability of conflict diamonds, is important as they have been a major source of funding for terrorism.[31] Additionally, the illegal gun trade and those diamonds and other gems are inextricably linked. The ability to control importation of illegal weapons directly impacts the safety of the civilians in the area, as well as U.S. SOF elements working there.

Oil is another commodity that has brought mixed blessings. As the price of oil has climbed on the global market, the potential revenue has increased dramatically. However, most of the oil currently being extracted is located on the west coast of Africa in an area known for instability. Thus, oil companies have become lucrative targets and a form of economically inspired irregular warfare has emerged. Oil is also found in Sudan, Angola, and several other countries. Rarely does it benefit the local population, as is covered in more detail later in the discussions of specific regions.

Fluctuating prices are problematic. It should be noted that rapid decline in oil prices, as occurred in late 2008, can have devastating effects on developing countries. Many of these oil producing nations failed to diversify their economies, falsely assuming that the price would always remain high (at least over $100 per barrel). Unpredicted decline in oil prices leads to further instability almost as much as dramatically inflated prices.

Uranium is abundant in areas of West Africa, which hold about one third of the earth's supply. However, foreign companies run the mines. The reluctance to share wealth derived from those mines has precipitated local conflicts. The efficacy, including safety issues, regarding the use of nuclear energy will be debated academically. From a pragmatic perspective, increasing the production from nuclear reactors is probably the fastest means to boost overall energy output. While that will not satiate the global energy demand, it will help. That means that uranium mining will likely increase, as will profits. How the money is distributed will impact local stability in areas with uranium bearing ore.

In addition, Africa has 90 percent of the world's cobalt as well as 90 percent of the platinum. Congo has extensive copper mines. As for chromium, 98 percent of the world's supply is on that continent along with 50 percent of the known gold reserves. Then, 64 percent of the world's deposits of manganese are located there. As in South America, deforestation is a grave concern. The Congo Basin is second only to the mighty Amazon for tropical forests. The demand for hardwood lumber is increasing worldwide and over-cutting is rampant on both continents. There are predictions that within 50 years this source could be depleted. Of note, six of every ten trees cut go to China.[32]

Natural resources are equally necessary for domestic consumption and viability. Africa has extreme agricultural diversity. The countries south of the Sahara frequently have very fertile lands while people living in or near the vast deserts have been continuously on the margin. Because of aggressive population growth in many countries, the upper limits of ecological sustainability are rapidly approaching, and in some cases have been passed. While there are currently about 680 million people living in Africa, that number is predicted to double by 2050. A recent World Wide Fund for Nature (WWF) report stated that Egypt, Libya, and Algeria are already living beyond their biocapacity.[33]

A recent World Wide Fund for Nature report stated that Egypt, Libya, and Algeria are already living beyond their biocapacity.

Egypt offers a classic example of the problem. For many generations the fertile Nile valley provided for the needs of the population. However, that constitutes less than 3 percent of the landmass of the country. The rest is consigned to the trackless Sahara desert with life supported only at a few sparse oases. For decades the Egyptians have worked aggressively to bring down their growth rate, and with some success as they transitioned from about 3 percent annually to the current 1.7 percent. However, the land is already overburdened and there is intense competition to build homes on agriculturally viable land while battling increasing desertification. There simply is not enough agricultural capacity to feed a growing population, even though Egypt exports a large labor force to other countries.[34]

A different kind of example can be seen in Rwanda. In this country the population growth rate is approximately 2.8 percent, which is very high.[35] When one enters the country it can be seen that every available area for farming already has been cleared. There are a few national forests that are home to the remaining, but endangered, mountain gorillas. With extensive publicity, these gorillas attract many foreign visitors who pay for the privilege of getting close to them in their natural habitat. However, there again is stiff competition for the forest. The local inhabitants are cutting the trees as much as possible to increase their farmland. As they do not directly benefit from the visitor's exorbitant entry fees, they do not share the government's concern for either protection of the land or safety of the gorillas.[36]

These competitive issues permeate the countries of the Lakes Region of Africa. Already noted was the Rwandan genocide of 1994. Despite public acknowledgement of gross inhumanity, the ethnic tension in the region today, especially outside Kigali, is palpable. The increasing population, coupled with declining food and other resources, could easily trigger another major outbreak of violence.

The other countries listed in the WWF report as pushing their limit on biocapacity include Morocco, Tunisia, Ethiopia, Kenya, Uganda, Senegal, Nigeria, South Africa, and Zimbabwe. Note that these countries are not concentrated in a single area but represent all sections of the continent.

As can be seen in these examples, Africa is a continent rich in natural resources, yet poverty is epidemic. To most Westerners, that situation is counterintuitive. For SOF operators it is important that they understand the realities of these situations, and the underlying principles that seem to cause frequent conflict. When intertribal conflict occurs, there is often a causal relationship to competition for natural resources.

Global Tolerance for Extreme Casualties

One of the most disturbing aspects of conflict in Africa is the tolerance for extreme levels of casualties exhibited by both African nations and the external world. At the end of World War II, when the extent of *The Holocaust* perpetrated by Nazi Germany became fully recognized, there was a collective hue and cry that stated *NEVER AGAIN*. Unfortunately, that sonorous commentary was specious and the world has repeatedly turned a blind eye toward genocide. Rwanda already has been discussed, but it is not the only example of when the world watched and waited. Starting in 17 April 1975, Pol Pot's Communist regime took over power in Cambodia and systematically executed or starved to death an estimated 2 million of its own citizens. People were killed for any reason, and ethnic/tribal differences were among them.[37] The United States, having just withdrawn from Vietnam, chose not to reengage in the region.

As the former Yugoslavia disintegrated into chaos there were widespread reports of ethnic cleansing and mass murders. The UN was slow to intervene and, even once troops were committed and observed atrocities, they were unable to halt the killings of thousands of people. Though debated by some sources, the international courts upheld the term of *genocide*. Estimates of fatalities range as high as 200,000 with two million refugees created.[38] Worth noting was how long it took any of the Western nations to intervene, even though atrocities were being widely documented.

The 1994 rampage in Rwanda that led to hundreds of thousands of deaths is not an isolated incident. In more than two decades of civil war between northern and southern Sudan, more than two million people died from violence, famine, and disease associated with that conflict. In addition more than 600,000 people from southern Sudan sought refuge in neighboring countries and one to two million more were internally displaced.[39]

Among the ongoing tragedies in Sudan is the conflict in the western region of that country known as Darfur. The problems were decades in the making. Prolonged drought, desertification, and overpopulation are among the primary reasons for the fighting, which breaks down along ethnic and tribal lines. The search for water pits the nomadic camel-herding Arab Abbala and Baggara tribes against the black African farmers.

Journalist, photographer, and former U.S. Marine, Brian Steidle valiantly documented the extent and nature of this disaster with his camera. He then presented his material to anyone who would listen, including members of the U.S. Congress.[40] While the U.S. has labeled the casualties as genocide, other governmental bodies choose to debate semantics but do little else. The United Nations estimates that since the conflict's inception in 2003, over 400,000 people have died either from attacks by the Janjaweed partisans who come from Arab-speaking tribes of northern Darfur, disease, or starvation. An additional 2.5 million people have been displaced. In response, the UN sent in 7,000 peacekeepers from the African Union, but they have proven to be ineffective at preventing continuation of the violence in Sudan and Chad. More peacekeepers have been authorized but have not been made available. The plight of Darfur has received some attention from various Hollywood personalities, but even their protestations have had only marginal effect. The complexities of this conflict are discussed in more detail starting on page 20.

Between South Sudan and Darfur the combined number of internally displaced persons in that country was greater than found anywhere else since World War II. Due to those conflicts, over four million people were forced from their homes. Some in South Sudan have returned home. That situation is still tenuous and is also addressed in the section on Sudan/Darfur-Chad starting on page 20. In response to these crises the U.S. has provided more than $4 billion in humanitarian, development, peacekeeping, and reconstruction assistance.

With over five million fatalities, the worst recent situation due to conflict is in the Democratic Republic of Congo. That too is covered later, in the section on the Congo starting on page 27. The point at this juncture is that it is possible for the world to ignore horrific casualties if they occur out of sight. Even if they are known, massive insults to humanity in poor countries with little political clout may continue for long periods with little outside intervention or support. Rarely have fatalities come from significant force-on-force combat. Most of these situations happen based on irregular warfare and the byproducts of those conflicts.

To this day tolerance for extreme casualties in Africa exists especially by the Western world. As SOF elements participate in operations on the continent, they may well encounter cognitive dissonance between their established American values and the situations in which they may find themselves. Therefore, it is essential that missions be chosen wisely, and troops prepared for the potential outcomes that are beyond their control.

Examples of Irregular Warfare In Africa

In many regions of Africa, conflict is a norm with historical precedence dating back for centuries. It is impossible to cover all of those conflicts, or to even address the current ones in complete detail. This section will acquaint the reader with some of the salient facts about ongoing and emerging violence, especially as they relate to American interests. Even cursory analysis of these situations clearly indicates that conventional U.S. military intervention would have extremely limited utility, except in very isolated cases. However, the missions for SOF elements abound, albeit they should be undertaken only with careful consideration and deliberations about the desired outcomes.

For any forces, including SOF elements, with assigned missions in Africa, it is important that they have a realistic understanding of the history and severity of the problems they may face. They, and their leadership, should know that even catastrophic levels of human suffering may not be sufficient to garner the external support necessary to resolve, or even mitigate, those circumstances. As demonstrated, Africans, and most of the people of rest of the world, are prepared to tolerate or ignore almost unimaginable levels of massive inhumanity. This certainly suggests that expectations must be carefully managed, including those of the troops involved. There are other examples of delays and denials in responding to conflicts and humanitarian crises. In fact, the worst example in modern history is ongoing in Africa today.

The Horn of Africa

The Horn of Africa is no stranger to war, and since the 16th Century the area now known as Somalia has experienced more than its fair share. Somalia was created from territory formerly dominated by the British and French colonialists and gained independence in 1960. At the same time the land previously controlled by Italy became Djibouti. A coup led by Major General Muhammad Siad Barre established a socialist government. However, continual fighting ensued, and in 1991 the central government of Somalia ceased to function. Civil war raged. In addition to tens of thousands of direct casualties, the scorched earth polices of the feuding factions brought hundreds of thousands of people to the brink of starvation, and many succumbed. Despite televised reports showing listless and dying children, the world simply watched and waited. Later that year the UN, supported by American forces, entered the country on what was designed to be a humanitarian mission.[41] What followed is well known in the annals of SOF history and heroism. The important lesson about events in Somalia at this juncture is how long the U.S. and the rest of the world waited before responding to the humanitarian crisis evolving before their eyes.

In 1994, without restoring stability, the United Nations Forces withdrew from Somalia, leaving local warlords ruling in a power-sharing agreement. In addition to internal instability, Somalia faced threats from Ethiopia, based partially on the existence of an ethnic Somali organization, the Oromo Liberation Front (OLF) that attempted secession of the Ogaden region in southern Ethiopia. Two major factions emerged in Somalia: the Transitional Federal Government (TFG), comprised of warlords, and the nonaligned Supreme Islamic Courts Union (ICU), comprised of 11 clans—many of which were former members of al-Ittihad al-Islamiya. Previously those members of the ICU had fought alongside of the OLF in support of their efforts to return the Ogaden to Somalia. While always restive, in June 2006, full-scale civil war had broken out and the ICU seized control of the capital, Mogadishu, and much of southern Somalia.

On 20 July 2006 thousands of Ethiopian troops entered Somalia to assist the TFG. By December the TFG forces were successful in regaining control of Mogadishu. Of concern to the U.S. was that the indigenous ICU forces reportedly had a substantial number of foreign mujahideen fighters assisting them. Some of these were believed to be members of Al Qaeda. In January 2007, based on intelligence reports that gave specific locations for Al Qaeda operatives, SOF AC-130 gunships joined the fight and conducted attacks.[42]

As the TFG, backed by Ethiopian troops, could win pitched battles, the mobile ICU adopted insurgency-based hit and run tactics. Collocating with their opponent, the insurgents would quickly launch mortar attacks, or ambush vehicles, and then melt back into the civilian population. The response led to cordoning off areas and returning fire, often with heavy weapons. This resulted in the deaths of many innocent civilians and the destruction of parts of Mogadishu. Even the introduction of African Union Mission peacekeepers did not stem the insurgency.

The fighting continued in 2008. Another U.S. missile strike on Dhusamareb was successful in eliminating senior insurgent leaders. Among them was Aden Hashi Ayro, a notorious Islamist terrorist.[43] However, the insurgency did not stop and there are increased reports of violence against civilians by all warring factions. By November Al Shabaab, a radical Islamist group with close ties to Al Qaeda, had taken over much of southern Somalia and threatened Mogadishu. Despite intervention by Ethiopia to stop the emergence of the hard-line Islamist courts, Al Shabaab is reported to be getting stronger, more sophisticated, and importing fighters from other parts of Africa.[44]

What has emerged from the continued violence is described as a "growing humanitarian crisis." All factions have vowed to continue fighting which is leading to deprivation for hundreds of thousands of civilians. No one knows how many people have already died as a direct or indirect consequence of this conflict. Low estimates place the number at 350,000 victims.[45]

Oil will be a recurring theme in this paper. Though not known as an oil producing country, exploration has begun in northern Somalia. The insurgents are concerned that if the Ethiopians are successful in finding oil, they will annex the land. High oil prices are making it worth the risk to conduct this exploration.[46]

Another international problem is a surge in piracy on the high seas just off the coast of Somalia. In 2008 there were numerous reports of pirates taking ships in the area and by fall, more than 120 ships had been attacked and at least 40 seized. Some of the more daring assaults catapulted the issue of piracy from relative obscurity onto the front pages of the news. Among those notable hijackings was the Ukrainian ship *MV Faina* with a cargo containing 33 Russian-built T-72 tanks and depleted uranium rounds destined for the Kenyan port of Mombasa. Probably the most audacious raid was the seizure of the 330 foot Saudi supertanker, *Sirius Star*, loaded with two million barrels of crude oil for which the pirates demanded $25 million ransom.[47]

Another international problem is a surge in piracy on the high seas just off the coast of Somalia.

The East African Seafarers' Assistance Program reported that "the pirates are making millions of dollars" by raiding ships and holding them for ransom. They indicated that this money is going to fund the conflict by both the insurgents and the opposing warlords.[48] Reports indicate that some of the pirates come on fast boats and are armed with modern weapons and that they use satellite phones and GPS equipment. The negotiation process can take months, and the Foreign Minister of Kenya estimated that these ventures net about $150 million annually, though some counterpiracy experts believe that number to be greatly inflated.[49]

In late 2008 and early 2009 the international news media placed considerable emphasis on the piracy problems in the Horn of Africa. In response, several nations deployed naval vessels to the area, which resulted in a sharp decrease in attacks. However, the respite was short-lived. By April the pirates had adapted to the patrols.[50] While maintaining their basic tactics, they merely expanded their area of operation. These pirates now were reported to be operating over 1000 kilometers from the coast of Somalia.[51]

There is little doubt that the conflict in Somalia will continue, probably indefinitely. American SOF elements have already been involved both as advisors and in limited direct action missions. The participants in the conflict come from many countries, and no one is immune from attack. Unfortunately, that includes aid workers who only come to attempt to ameliorate an already desperate situation. Somalia provides an easel on which is portrayed the complexities of irregular warfare and just how adaptable the adversary will be.

Sudan/Darfur and Chad

Sudan is a vast country that covers more that one million square miles. In recent years, the world's relationship with Sudan has largely focus on one area, Darfur. The plight of Darfur, and the ongoing genocide there, was mentioned earlier. In reality, that conflict is far broader and more complex than just the Janjaweed bandits' attacks on civilians. Darfur is in the western region of Sudan; the conflict is again between ethnic and tribal groups that live in the area. Like many African conflicts, there are centuries of contentious history leading up to the current situation. As with the Tuaregs in the western Sahel, tensions often arose between nomadic and stationary people, especially when water was scarce.

The current rebellion began in 2003 when two rebel groups, the Sudanese Liberation Army/Movement (SLA/M) and the Justice and Equality Movement (JEM), accused the government of Sudan of favoring Arabs and oppressing non-Arabs (sometimes identified as Africans). While supported by the Army of Sudan, the primary counterinsurgency forces used against the rebels were the Janjaweed, an armed group of partisans from Arab-speaking tribes of northern Darfur. The Janjaweed are known locally as "the devils on horseback," though these highly mobile forces also ride camels and have access to pickup trucks.

While attempting to counter the SLA/M and JEM forces, the tactics employed by the Janjaweed have been extremely brutal. They are known for surrounding villages, then killing everyone they can capture and burning the huts to the ground. These attacks were sometimes supported by aerial bombardment by Sudanese aircraft. It has been alleged that Russian pilots have flown at least some of those missions.[52,53] That claim is refuted by Russia, along with their denial of breaking the arms embargo by supplying weapons to Sudan. China is a major weapons supplier to Sudan. In addition, they have been involved in training fighter pilots for A5 Fantan fighter jets that were delivered before the embargo was enacted.[54]

As indicated earlier, the estimated number of fatalities runs as high as 400,000, though the Sudanese government places the number much lower. In addition to the deaths, an estimated 2.5 million people have been displaced.[55] Despite repeated denials of his active support for the Janjaweed, in July 2008 the International Criminal Court (ICC) filed genocide charges against the President of Sudan, Omar al-Bashir. The ten counts include charges of organizing a campaign of murder, rape, and deportation against the civilians living in Darfur. Al-Bashir has hidden these crimes under the guise of "a counterinsurgency strategy and unrelated intertribal clashes." The prosecutor noted that al-Bashir's intention was "to end the history of the Fur, Masalit, and Zaghawa people."[56]

On 4 March 2009, after much deliberation, the ICC took the next step and issued an arrest warrant for al-Bashir.[57] This is the first such warrant issued for a sitting head of state and it has complicated the situation throughout the region. One of al-Bashir's immediate responses was to expel many of the non-governmental organization (NGO) aid workers who were providing assistance to the victims of violence in Sudan.[58] In addition, Arab leaders have failed to enforce the warrant as al-Bashir has traveled to several countries in the region. At the 21st Arab League summit those leaders issued a communiqué stating they believe the warrant to be a violation of the Vienna agreement of 1961.[59] This snub of the ICC is a problematic precedent for future criminal indictments they might attempt in the region or elsewhere in the world.

The attacks have not stopped. Even after people have been displaced and are living in refugee camps, they remain in danger. In June 2008 the police commissioner of UNAMID, the inadequate United Nations peacekeeping force, stated that some of "the biggest issues are the rapes and crimes against women and children." The Janjaweed have used extremely brutal rape as a systematic weapon in ethnic cleansing of the area. There are female victims as young as four years of age who have been reported. The female victims have indicated that the Janjaweed make no attempt to hide their rapes, which often occur in front of family members. Male relatives rarely object as they are instantly killed for interfering. Gang rape is prevalent and sometimes leads to pregnancy. In fact, there is now a problem of babies from these rapes being abandoned by the mothers.[60]

More foreign aid workers are aware of the problems, but state they fear making reports, as the Sudanese Government will have them deported. On the record, Mohammad Hassan Awad, a Sudanese Humanitarian Aid Commissioner for West Darfur, stated, "There is no rape in Darfur." Contrary to overwhelming physical evidence, he claims that the foreign aid workers make up these stories.

For several years there were camps for *internally displaced persons* (IDPs) in Sudan that were run by humanitarian groups. While there was agreement that the refugees would be secure when in those camps, that turned out not to be the case and raids continued. In many cases, civilians were forced from their villages into IDP camps, only to be chased from the camps into the bush where many of them died.[61]

This conflict is not confined to the Darfur region of Sudan. In order to escape from the terror of the Janjaweed, hundreds of thousands of refugees crossed the border into Chad where camps were set up. While once believed immune to attack, raids into Chad became commonplace, and neither refugees nor aid workers were safe.

Again, there is a long history involved in this episode. There was a lengthy conflict between Chad and Libya in the 1970s and 1980s. The latter part was known as the *Toyota War* memorializing the seemingly ubiquitous pickup trucks driven across the desert carrying missiles. The war finally ended when the Chadian Army, backed by the U.S. and France, stopped Qadaffi's forces. However, during that conflict, Qadaffi had armed rebel forces in Chad. Some of these armed rebels later became known as the Janjaweed.

In 2005, incited by a questionable change in the constitution of Chad that altered term limits for the president, armed resistance groups began another insurgency. In addition to the Sudanese-backed Janjaweed, there were native insurgents from Chad as well. These included a number of senior officers who defected from the Chadian Army and took some heavy weapons with them.

Significant fighting occurred in the towns that bordered Sudan. In 2005 through 2007 despite a truce, there were incursions by all of the parties involved in the conflict. Caught in the middle were the refugees and their humanitarian aid workers. There were frequent reports of humanitarian aid vehicles being ambushed and the contents stolen in both Darfur and Chad, making a difficult situation far worse. The World Food Program reported that 130 of their trucks were hijacked en route to the refugee camps in 2007. In 2008, that number of hijacked trucks was exceeded by June.[62]

In February 2008, despite the presence of some European Union (EU) peacekeepers, the native rebels, with support from Khartoum, began a major attack against Chad. Until this time, irregular warfare, with hit and run tactics, had been the norm. This attack saw thousands of fighters in hundreds of vehicles sweeping across Chad. In three days they had entered the capitol city of N'Djamena, located in the southwest corner of Chad, over 500 miles from the Sudanese border.[63]

This operation demonstrated how quickly irregular warfare can be transformed into a larger context. The presence of 3700 EU soldiers did not halt the attack, as their mission was to protect the refugees. It also indicates how little regard for the presence of the EU was shown by the Khartoum government. Despite historic security relationships between France and the countries of North Africa, Sudan assumed that they would sit on the sidelines unless directly confronted. In fact, the main mission of the French soldiers was to protect foreigners in N'Djamena, and prepare to evacuate them if necessary.[64]

Although the rebel forces withdrew from N'Djamena, they have continued fighting in other sections of the country. Of the 26,000 peacekeepers authorized by the United Nations, only about 7,000 have entered the area. The EU contingent is relatively well-equipped, organized and led. The remainder of the troops come from the African Union (AU) and are poorly equipped and trained. Most importantly, the number of peacekeepers is totally inadequate to maintain control over the vast areas of Chad and Darfur that are being contested. In addition, even the peacekeepers have not been immune from attacks and robberies. As an example, in September 2007, rebels attacked an AU peacekeeping camp in central Darfur. They killed 10 soldiers and stole heavy weapons. The peacekeepers usually do not fight back when attacked. Major General Martin Luther Agwai of Nigeria, the senior mission commander of the hybrid force, indicated that any use of force would be seen as taking sides and he might lose cooperation from the Sudanese Government. He stated, "I'm not here to stop bullets."[65]

There are other problems with the peacekeeping units from the African Union. The quality of troops is very inconsistent. Equipment often comes without necessary logistical support to keep it running. In addition, it may be an item that the troops have never been trained to use. Weapons are poorly maintained. Even pay can be a problem. A Norwegian major, on assignment as a UN staff officer, noted that some AU peacekeepers he observed had not been paid in as long as 10 months while he was there in 2007.[66]

Besides intertribal and interracial tensions, there are several issues at the heart of the Sudan-Chad situation. Oil is a key factor, as both countries have oil. Sudan has established relationships with Russia and China that include shipment of oil. In return, both China and Russia have provided arms to Sudan, in spite of the UN embargo. In addition, both countries have blocked measures designed to sanction Sudan when they have been brought before the Security Council.

Chad also has recently discovered oil. It is moved via pipeline to Cameroon and then offshore to waiting tankers en route to the U.S. Chad is one of the five poorest countries in the world with about 78 percent of its people living on less than one dollar a day. However, the profits from oil have not benefited them, but rather go into the external bank accounts of the power elite of the country, and to foreign oil companies.[67]

The current Sudan/Darfur-Chad predicament represents irregular warfare at its worst and most complex. The situation is likened to the Wild West days in American history. Banditry is widespread and there are dozens of rebel factions competing for the spoils. The rebel groups are complemented by militias comprised of tribe members and former government soldiers. The UN has been unable to fulfill the authorization for 26,000 peacekeepers, which is probably the bare minimum necessary, given the size of the territory and number of IDPs.

The current Sudan/Darfur-Chad predicament represents irregular warfare at its worst and most complex.

The president of Sudan, al-Bashir, has gone on record stating there "is an influential power that does not want peace in Darfur." [68] China is the most likely suspect as they are benefiting greatly from their trade agreements and illegal sale of weapons. Another complicating factor is the economic relationship between United Arab Emirates (UAE) and Sudan. After China, the UAE is the second largest investor in Sudan with a total of about $7 billion. While the relations between the U.S and Sudan are strained, the UAE is considered a close ally of America in global counterterrorism operations.[69]

Further, all of the conflicts in Sudan are not located in the Darfur region. As indicated earlier, a civil war raged in Sudan for over 20 years resulting in over 2 million deaths, and millions of displaced persons. At war were the Government of Sudan, based in Khartoum, and the southern forces known as the Sudanese People's Liberation Movement/Army (SPLM) and located in Juba. Finally, in 2005 the two sides signed the Comprehensive Peace Agreement (CPA) that called for power sharing, but included a provision under which southern Sudan would have a referendum on its independence.[70]

Unfortunately, many of the provisions of the CPA have not been implemented. For a while, the SPLM withdrew from the agreement because of these issues, which include a lack of money promised for oil production. There are few guarantees that the CPA commitments will be met and indicators for continued peace are not good. There is high risk of renewed fighting in the oil-production region of Abyei, which is in southern Sudan.[71] While China is supplying Khartoum with weapons, the SPLM in Juba is arming and training for war as well. An anonymous Canadian UN observer is quoted in the *Christian Science Monitor* as saying; "It's a lull in which both sides are regrouping for the new war."[72]

Andrew Natsios, who served as a U.S. Special Envoy to Sudan in 2006–2007, has recently written about an impending crisis. In a lengthy and compelling article in *Foreign Affairs*, he stated, "Trends more ominous than even the carnage in Darfur could bring the country far more bloodshed soon."[73] The referendum on independence is guaranteed by 2011. Polls indicate that over 90 percent of the population of southern Sudan is in favor of separation. Both sides have relatively large armies but more frequently employ partisans, like the Janjaweed, as a counterinsurgency strategy. Neither side has strong control over the forces that work for them. Natsios also quotes an unnamed, highly respected African diplomat saying, "If the north and south (Sudan) return to war, it will unlock the gates of hell."[74]

The government in Khartoum is not given credit for strategic thinking. Natsios states, "They are prepared to kill anyone, suffer massive civilian casualties, and violate every international norm of human rights to stay in power, no matter the international pressure, because they worry (correctly) that if they are removed from power, they will face retaliation at home and war crimes trials abroad."[75] Talks are underway by the SPLM to unite 27 rebel groups in Darfur, and other marginalized groups, against the Sudanese government. If the CPA collapses, it is likely that Africa will experience irregular warfare on an unparalleled scale.

Natsios persuasively argues that no Western country, including the U.S., is likely to invade Sudan, even if casualties mount. There is however, another element that will have emotional impact of such decisions. While we are a secular nation, religion does matter in political decisions. While there are animists in South Sudan, the majority of the population are Christians. The forces of the north are comprised of Islamic Arabs. While religion has often been a factor in conflicts in Africa, the emerging situation represents that worst of all possibilities—a racial and religious war that determines who controls the oil fields.

In early 2009, this complex *perfect storm* for conflict became even worse. As noted earlier, the Chinese are already deeply involved in the affairs of Sudan—oil being their primary interest. Then, rather unexpectedly, Israel entered the equation bringing the Middle East problems directly to this region. On 27 December 2008, Israel had initiated an attack against Gaza. The stated intent of the incursion was to stop the missile attacks that had been launched repeatedly from that Palestinian territory. One of the primary complaints of Israel was that significant amounts of arms were being smuggled into Gaza through their border with Egypt.

Israeli intelligence indicated that Iran's Islamic Revolutionary Guards were sending arms through Sudan and they had specific knowledge of a shipment.[76] The 23-truck convoy was struck by dozens of fighters in two bombing runs as they crossed the Sudanese desert headed for the Egyptian border. Arabic newspaper sources reported that U.S. officials previously had warned Sudan to stop allowing Hamas to conduct smuggling though their territory.[77] Sudan did not act on the advice.

Farther to the south there is Uganda, and another set of issues that are sometimes related to those across their northern border.

Uganda

Southern Uganda, along the northern edge of Lake Victoria, has made significant advances since the vicious dictatorial reign of Idi Amin. The capital city of Kampala is relatively modern and the road system is qualitatively better than that of neighboring Kenya. Stability is such that many Europeans vacation in the area and Entebbe is a tourist site. However, there is another area in Uganda that remains restive.

Among the little-known conflicts is the long-enduring fight in northern Uganda. There, the Lord's Resistance Army (LRA) has been attacking the Ugandan Army since 1987. Paradoxically, this group, known for extreme violence including killing, rape and large-scale child abductions, claims it wishes to establish a theocracy based on the Biblical Ten Commandments. Despite repeated peace agreements, the fighting has continued albeit at lower levels.

Most rebels are members of the Acholi tribe, which populates both northern Uganda and southern Sudan. Though called a civil war, several international boundaries are involved, and the LRA was believed to have been supported by bases located in Sudan for many years. Since an agreement between Uganda and Sudan in 1999, violence has eased. However, the LRA also operates from the nearby Democratic Republic of Congo (DRC). Joseph Kony, the LRA leader, is a war criminal wanted by the International Criminal Court.[78] The numbers of victims of this war are not clear. There are estimates that over 20,000 people have been killed, and up to 2 million displaced. Most of those came from the Gulu and Kitgum districts. It is believed that the LRA is still holding at least 3,000 kidnapped children, many of whom they force into combat.[79]

While the Ugandan military has put pressure on the LRA, the rebels have proven to be a resilient force. Reports indicate that resupply sometimes is being accomplished via air drops in the mountainous regions of Karago. The conflict in Uganda is tied to the Sudanese conflict previously discussed. The southern Sudan leadership believes that the LRA logistical support originates in Khartoum. The LRA operations are not restricted to Uganda and continue to displace large numbers of civilians in both Congo and Sudan.[80]

There have been periodic peace negotiations between the Ugandan Government and the LRA yielding many promising news releases of imminent accords. Yet, they never quite get signed. In June 2008, multinational plans to attempt to crush the LRA were drafted by the armies of Uganda, Sudan, and the DRC. In August it seemed resumption of the hostilities was likely.[81] Late in 2008 a multinational effort chased Kony and his forces. The result became known as the Christmas Massacre. United Nations sources claim between 200 and 400 people were hacked to death by the retreating LRA.[82] This is an example of irregular warfare conducted by a relatively small force causing greatly disproportional consequences to innocent inhabitants of the area. Illogical borders and ethnic strife exacerbate the situation.

Congo

The Democratic Republic of Congo (DRC, also DR Congo) has a very complex social milieu consisting of over 200 ethnic groups spread over a vast area that is about one-quarter the landmass of the United States. There is no majority group and there are many distinctly different languages spoken.

The largest ethnic groups include the Luba, Kongo, and Anamongo. Both Tutsis and Hutus have a significant presence as well. As a former Belgian colony, French is the official language used by the government. Four indigenous languages are officially sanctioned, Kiswahili, Lingala, Kikongo, and Tshiluba. Throughout its history, social discrimination based on ethnicity has been widely practiced and accepted.[83]

The area is rich in natural resources, which have been at the core of several conflicts. Congo has some of the largest gold, copper, diamond, and tin deposits in the world. Also of concern are the uranium deposits. It was the Shinkolobwe mine in Congo that provided the uranium for the Atomic bombs dropped on Hiroshima and Nagasaki.[84] More recently, illegal mining has taken place there with allegations that the ore was provided to Iran.[85] The borders are ill-defined and, like many other African countries, were imposed by European occupiers long ago. The natural resources available in Eastern Congo have caused near-constant conflicts as various groups vie for control of the area.[86]

Flying below the world's political radar has been the long enduring, and ongoing, war in Congo, which for a short time was called Zaire. Recent estimates claim that as many as 5.4 million people have died in the past decade as a direct, or indirect, result of that conflict. The fighting, formally known as the Second Congo War, involved eight nations (DR Congo, Namibia, Zimbabwe, Angola, Uganda, Rwanda, and Burundi) and about 25 separate armed groups and militias. While the war officially ended in July 2003, hostilities still continue.[87]

Ignobly, the Second Congo War holds the highest death toll of any conflict since World War II. It is estimated that about 45,000 people still die every month from violence, disease, and malnutrition, all of which are attributable to this conflict.[88] With the possible exception of periodic television specials, this tragedy goes almost unnoticed by the media organizations of the world. The number of victims and the scope of brutality are staggering. Use of ethnic cleansing was pervasive. There are claims that not only were a minority group, the Pygmies, being systematically exterminated, they were actually being cannibalized as well. Many Congolese see the Pygmies as *subhuman* and believe their flesh contains magical powers.[89]

Especially hard hit of all of the Congo War victims are children under five years of age. While they comprise 19 percent of the population, they account for 50 percent of the deaths. The leading causes of these juvenile deaths are preventable diseases and malnourishment. However, a substantial number have fallen victim to machetes.

While rape in warfare is not new, the application of systematic gender-based violence in the Great Lakes region of Africa has been especially vicious. Not only are women repeatedly gang raped, they are often mutilated as well. The attackers are quite aware that the damage inflicted goes far beyond just the female victim. It encompasses the entire village. Social mores of most Congolese cultures dictate that a woman who has been violated becomes an outcast to the community, even though she is a victim of a heinous crime. Amnesty International reported in 2004 that, "All the armed forces in the DRC conflict have committed rape and sexual violence, including the government armed forces of DRC, Rwanda, Burundi, and Uganda."[90] Some soldiers boldly state that they believe raping women will lead them to success on the battlefield. There are at least 200,000 known rape victims in Congo and recent reports indicate that these attacks are continuing.[91] Use of these tactics, especially when condoned by government military forces, is a major impediment to restoration of stability. The devastating effects of the crime can never be overcome.[92]

... the application of systematic gender-based violence in the Great Lakes region of Africa has been especially vicious.

In 2007, the first survey was conducted in an attempt to understand the magnitude of the problem. It was found that the civilians feared government soldiers as much as the rampant militias. A huge number, 55 percent, reported they had either been forced to work or been enslaved by armed groups. At least 34 percent had been abducted for more than a week, and 23 percent stated they had seen acts of sexual violence. In fact, 16 percent indicated they had been the victims of a sexual attack.[93]

The conflicts in Congo, and the rest of the Great Lakes region, represent the complexities of irregular warfare. The adversaries are ill defined and often morph from one loosely-knit coalition to another based on current exigencies. Support sometimes includes accessing assistance from military units in neighboring countries. Tribal affiliations play a role is such interventions. In general, there are rarely pitched battles or front lines. Rather, militias engage in periodic confrontations that minimize the possibility of their forces being pinned down and wiped out. Military units are located to protect key facilities, resources and infrastructure including airfields, ports, and mines. Troop discipline, even among formal armies, is poor. All elements, including members of the armed forces, engaged in the conflict have been accused of rape, pillage, looting, and random murders. These actions are often premeditated actions for ethnic cleansing. Even the foreign UN peacekeepers have been accused of sexual exploitation and abuse.[94]

In recent years, Congo has maintained only a fragile stability through intervention of 17,000 UN peacekeepers. While an election has been held, there is no guarantee that the country can be held together. In January 2008 *CBS News* reported that, "Fighting has broken out once again in eastern Congo and the region threatens to slip into all-out war."[95] One of the key personalities is a renegade Congolese Tutsi, General Laurent Nkunda. The minority Tutsis claim to be Congolese, but are often excluded because of ethnicity. As a remnant of the Rwandan crisis, conflict with militias is nearly constant and Nkunda's forces often clash with the Army.[96] In late August 2008 the UN peacekeeping mission confirmed that fighting between the Army and Nkunda was continuing.[97]

During editing of this monograph the situation in Congo deteriorated even further and the fierce fighting was characterized as a "human catastrophe" by the International Red Cross.[98] Ceasefires came and went. In November, General Nkunda's forces occupied several key towns in Eastern Congo and fear of increased foreign intervention from Angola, Zimbabwe and Rwanda fueled an already impossible situation. In addition to an escalation in civilian casualties, tens of thousands of people were again displaced with aid workers reporting extreme difficulty in providing assistance because the routes are often cut off. The aid workers reported that over 100,000 refugees north of the key city of Goma were without support.[99]

On 22 January 2009 General Nkunda was arrested after crossing into Rwanda. This was partially brought about by rapidly changing political considerations between Rwanda and DR Congo and the conduct of relatively successful joint operations. Given the prior support General Nkunda received from Rwanda, many observers are still concerned about the real implications of the arrest. There are many troops in the area who remain loyal to him.[100] Almost predictably, once the operations ended and some of the troops withdrawn, the militias reemerged and were again terrorizing the local population.[101] Unfortunately, the situation in Congo, and throughout the Great Lakes Region, is not likely to get better in the foreseeable future.

The Niger Delta and MEND

The Niger Delta, with an average rainfall of 157 inches along the coast, in one of the wettest places in Africa. It is an oil-rich region with a complicated history of violence and exploitation. The Niger River Delta is comprised of about 5,600 square miles of mangrove swamps and dense vegetation crisscrossed by small streams and creeks, making travel by land almost impossible near the coast. However, the major rivers, the Niger, Benin, and Imo, are navigable by fairly large vessels. Throughout the area, small boats are the normal means of transportation.[102]

There are an estimated 34 billion barrels of oil beneath this region, which is populated by some of the poorest people in Africa. While some efforts to steal oil are professional and organized, many Nigerian people are so poor that they are willing to risk their lives to pilfer a bucket of oil from the pipelines. Drilling small holes to extract oil has caused fatal explosions on several occasions. Despite the risks, the people return.[103]

The post-colonial history of Nigeria has been quite tumultuous. From the time of independence from the United Kingdom on 1 October 1960, the political parties largely broke down along tribal and religious lines. For the next 5 years the members of the Igbo and Yorubas jockeyed for power resulting in weak governance. Disequilibrium and corruption led to a series of military coups beginning in 1966. In May 1967 the Eastern Region declared independence calling the country Biafra. That action led to a 30-month civil war.[104] The consequences of that conflict were disastrous, with between one and three million people starving to death, and it ended when Biafra surrendered. Charges of genocide were leveled, but not substantiated due to the legal determination that the deaths were simply from famine, and not intentional extermination based on ethnicity.

Nigeria joined OPEC in the early 1970s as oil became increasingly profitable. While the military-led governments received large amounts of money for the oil, much of that was siphoned off because of epidemic corruption. Unfortunately, the government allowed the foreign oil companies a relatively free hand in drilling and failed to manage the process effectively. They also failed to diversify the economy and concentrated all of their efforts on the lucrative petroleum industry to the detriment of establishing domestic stability.

Since 1956 foreign oil companies have been drilling in the Niger Delta area. The reserves were nationalized in 1971 and placed under the state-owned Nigerian National Petroleum Corporation. By 2005, profits to that company had grown to over $60 billion. However, there are estimates that over 70 percent of the oil money has been stolen, in a country that ranks 142 of 163 on the Corruption Perception Index maintained by Transparency International. Major Brian Lionberger, in his master's thesis at Fort Leavenworth, Kansas, explored in depth the requirements for counterinsurgency in the Niger River Delta. Among the prominent issues of concern, he noted that corruption is so rampant that "it extends to every aspect of Nigerian political and social life, with every process of determination, selection, equity, and justice being compromised by influence and intimidation."[105]

Despite immense wealth from natural resources, most of the 30 million people living in the Niger Delta do not benefit from it. It is estimated that the vast majority of the population exist on less than one dollar a day.[106] Making matters worse is that the drilling operations showed little concern for preserving the local environment, and the oil industry infrastructure seriously disrupted the meager means for the local inhabitants to fish these waters.

Nigeria, a member of OPEC, is the United States' fifth leading supplier of oil, and in 2002 the U.S. State Department declared the area to be a strategic national interest.[107] Royal Dutch Shell has thousands of wells in 90 oil fields. In addition, the pipelines required to transport the oil extend over 3,700 miles and safeguarding facilities is problematic. The Nigerian military is ill prepared to provide the protection necessary to keep the oil fields at full production. Not only are they are inadequately trained and equipped, the epidemic corruption includes members at all ranks. The same is true for the police forces of the country.[108]

Early in 2006, out of the intractable milieu of poverty, corruption, and violence emerged a loose-knit, but determined, confederation of indigenous people who would challenge both the government and foreign industrial giants. Known as the Movement for the Emancipation of the Niger Delta, or MEND, they ostensibly dedicated themselves in an armed struggle against the exploitation of the local population. One of the MEND leaders with a nom de guerre, Major General Godswill Tamuno, told the BBC that their goal was to obtain total control of the Niger Delta's oil wealth.[109]

The actual make-up of MEND remains unknown. At least one analyst describes MEND as more of an "idea" than an established organization. At best MEND is comprised of several subelements, and can change names whenever that is advantageous.[110] This type of functioning was predicted and is consistent with the *superorganism* concept of terrorist and insurgent organizations put forth by the author in the first JSOU paper for the National Intelligence Council.[111]

The number of active insurgents ranges from a few hundred to several thousand. They are estimated to have over 100 boats and thousands of small arms. Their tactics include destruction of pipelines, kidnappings, and raids conducted at the time and place of their choosing. They have demonstrated the capability to conduct multiple coordinated attacks as shown on 28 January 2007 when, operating from speedboats, they successfully freed an imprisoned leader, Sobomabo George, in the center of the principal city of Port Harcourt. At the same time, the force of over 50 insurgents conducted diversionary attacks that resulted in the release of 125 other prisoners.[112] The damage to the oil infrastructure resulted in a reduction in oil output of approximately 500,000 barrels per day.

On 20 June 2008 MEND was able to shut down 10 percent of Nigeria's oil production with a single attack. This took place on the Shell-operated Bonga oil platform which was assumed to be out of reach of the MEND forces. Attacking this platform located 120 kilometers off shore demonstrated sophistication not previously observed and proved that all of Nigeria's oil production is within range of MEND.[113]

It is important to note that although the number of active insurgents in MEND is relatively small, they have the ability to cause global consequences. Due to the fragility and sensitivity of the petroleum market to all adverse events, even isolated attacks on platforms in the Niger Delta will cause the price of oil to increase substantially, often several dollars per barrel in a single day. The impact on Nigeria has been to decrease revenue from oil exports by about $1 billion a month.[114]

By some accounts, MEND is becoming increasingly sophisticated in its political activities. In addition to demanding access to the oil revenue, they have also demanded Shell pay $1.5 billion in compensation for polluting the Niger Delta. There are also indications that MEND may envision itself as a legitimate political party, and thus able to negotiate with the Nigerian government directly. Another of their demands is that the Nigerian military forces be withdrawn from this area. The ability of local forces to conduct counterinsurgency operations in the Niger River Delta is at best marginal.

The physical and political infrastructure is weak. The widespread perception by the indigenous people that the government is corrupt at all levels has provided substantial support for the MEND agenda.

Turmoil continues in this oil-rich area. In the first three months of 2009 nearly 130 people were kidnapped. These included both Nigerian citizens and foreign workers. While most victims have been released alive, this is a constant threat for the oil companies.[115] At the same time Chevron has confirmed continued attacks against their production facilities which caused them to stop pumping significant amounts of oil.[116]

In addition to the attacks, a great amount of oil is being stolen. One estimate 2009 put the thefts at about 680,000 barrels per day. While some thefts are by individuals willing to risk tapping a pipeline for a few buckets, there are also thefts on a wholesale level. Of deep concern is that some of this allegedly has been traced to retired senior general officers. This appears to be sufficiently blatant that militants were surprised that security officials even asked about their relationship to these criminal activities.[117] The amount of money lost because of oil thefts, shutdowns, and spillage in 2008 was estimated at $20.7 billion.[118]

The Niger River Delta will continue to be a vital interest of the United States. However, the U.S. is far from alone in these concerns. It has been predominantly European companies that have extracted the Nigerian oil. The Chinese have been investing in the industry, and like the Western companies, they too have been warned by MEND about the dangers that exist for all foreign workers. Support for counterinsurgency by U.S. SOF elements will be at best, extremely difficult.

The Tuaregs

Nigeria has other insurgency problems, in addition to those in the Niger River Delta. In the arid Sahel to the north, and also spanning parts of Mali, Niger, Algeria, Libya, and Burkina Faso, live the nomadic Tuaregs, known popularly as the *blue men of the desert* for the often blue-colored veils worn by the men. An ethnic offshoot of the Berbers, the Tuaregs have been fiercely independent for centuries and thus have come into periodic conflict with various governments. Somewhat victims of the boundaries established long ago by Europeans, the Tuaregs have been reduced to minority status and marginalized in every country in which they currently reside.[119]

The territory they cover is vast, yet extremely sparsely populated, and generally inhospitable to human habitation. Life for the Tuaregs has always been tough, and yet many still prefer to live in tents in the desert. As I observed in Timbuktu, Mali, poverty is at such a level that it is only the middle class that can move up to mud huts. Negotiations with the Tuaregs can be formidable, and they go to lengths to differentiate themselves from other indigenous people.[120]

General repression of the Tuaregs was exacerbated by repeated droughts in the 1970s and 1980s, which forced changes in traditional migration routes. Those changes brought the Tuaregs into conflict with other groups that lived in more stationary, even urbanized, settings. Those droughts also weakened the central governments of several countries. Barely able to cope with local economic problems, the governments did little to assist in the plight of the Tuaregs. This led to a movement by some Tuareg leaders to call for autonomy. To that end, the Popular front for the Liberation of Niger (FPLN) was formed. In May 1990, with grievances escalating, the FPLN attacked a police station in Tchin-Tabaradene. In response the Nigerien military arrested, tortured, and killed several hundred Tuareg civilians there and in nearby cities in what became known as the Tchin-Tabaradene massacre and led to more than five years of open conflict.[121] The conflict expanded to Mali and continued until peace accords were signed in 1995.[122]

Above, a Tuareg on the move; below, close-up of a Tuareg guide. Photos courtesy Lt Col Rudy Atallah, U.S. Air Force.

Despite the ceremonial burning of weapons at Timbuktu in 1996, the peace was an uneasy one, punctuated periodically by isolated raids. Then, in February 2007, the Second Tuareg Rebellion began with attacks by the Niger Movement for Justice (NMJ) with the intention of gaining greater economic development. Like MEND, the Tuaregs feel that they have been left out of the benefits of natural resources in their territory, specifically uranium. In fact, the generally impoverished Niger holds some of the richest uranium deposits in the world and amounts to about 72 percent of their national export income. Like oil in the Niger River Delta, very little of the substantial profits from the mines go to the local inhabitants.

Any attack in the uranium-mining region has immediate international impact. The French nuclear power industry is dependent on the Arlit mines. The Chinese company, China Nuclear International Uranium Corporation (SinoU), also has a deal to mine uranium at Teguida, which is in a territory used for grazing in the winter months.

By September 2007, much of the fighting shifted from Niger to Mali. At the end of August there were several attacks against both military and civilian targets. In September it was reported that the Tuareg fired on, and hit, a U.S. Air Force C-130 that was dropping supplies to a Malian Army unit near Tin-Zaouatene.[123]

In 2008, despite efforts to negotiate a peaceful settlement, attacks continued. During July rebels attacked a remote paramilitary outpost near Tessalit, in the Kidal region of northeast Mali, capturing the entire contingent.[124] In late August in northern Mali, Tuareg insurgents seized a military supply convoy and took 23 soldiers hostage. The following day there was a second attack involving an ambush and a gun battle that lasted several hours. Again, large quantities of munitions were reported taken in the raid.[125]

For several years, members of U.S. Army Special Forces Groups have been conducting training missions in the region. The effectiveness of that training is not known, but they were clearly hampered by lack of adequate linguistic capability.[126] Of some concern is the loyalty of those forces trained by U.S. SOF units. In August 2007, the NMJ claimed that the entire Niger Rapid Intervention Company, recently trained to conduct counterterror operations by U.S. Special Forces, had defected to them en masse. While that report was uncorroborated, it is known that Major Kindo Zada, a close associate of the murdered former president, did defect and arrived with twenty vehicles full of soldiers.[127] Such actions raise a significant issue for SOF trainers. What are the loyalties of troops sent to them, and are they ever vetted before training? There will be no perfect answer, but such defections raise serious questions.

In March 2009 Libyan leader Muammar Gaddafi sponsored a peace agreement between the Tuaregs and the and the government of Niger. Following negotiations, a comprehensive peace settlement was announced. There have been several such agreements in the past, and they have proven to hold only temporarily. One unresolved issue at the time of publication has been Europeans who were kidnapped and are still being held hostage. Many believe they were taken by the Tuaregs but later turned over to elements of Al Qaeda in the Islamic Maghreb that operate in the same areas.[128]

Al Qaeda in the Islamic Maghreb

In recent years another transnational terrorist organization has emerged in the Sahel. In January 2007, from the shadows of earlier insurgent organizations, came Al Qaeda in the Islamic Maghreb (AQIM). Elements of

A Malian boy watches incoming AFSOC CV-22 Ospreys carrying troops from Mali and Senegal during Exercise Flintlock '08. The exercise provides USSOCOM forces the opportunity to work with African and European partner nations in the effort to build bonds of trust and confidence. U.S. Air Force photo/Capt Bryan Purtell.

AQIM have been active in several countries in Northern Africa including Algeria, Mali, Niger, Mauritania, Libya, Tunisia, Morocco, Senegal, and Chad. The complex history of the development of terrorist activities in the North Africa Maghreb countries of Morocco, Algeria, and Tunisia are discussed comprehensively in the book, *Terrorism in the Maghreb*, by Anneli Botha, a counterterrorism expert from South Africa. She notes that there are important differences in how each country has responded.[129] We agree that this is a growing threat and not limited to this area, as these terrorists have exported their craft to Europe, and that this is a topic of grave concern throughout that continent.

Prior to assuming the AQIM name, this Sunni Muslim jihadist group was known as the Salafist Group for Preaching and Combat (GSPC) and had taken responsibility for numerous terrorist attacks in the region. However, some analysts believe that more than a name change was made and that inclusion of the Al Qaeda brand signaled widening ambitions. With formal cooperation with Al Qaeda, AQIM acquired better financing, became more sophisticated, and was able to pursue global objectives in addition to their local attacks.[130]

Arab recruitment of jihadists in North Africa can be traced to the late 1970s. Hundreds of volunteers went to Afghanistan to fight against the Soviets. At the conclusion of the conflict they returned to their homes, bringing their fighting skills and experience with them. Historically, AQIM is related to other previous groups including, the Armed Islamic Group, known by the French acronym GIA. During the 1990s GIA vowed to destroy the secular government controlling Algeria. They employed violent terrorist tactics, which included beheadings, against intellectuals and journalists. Showing their animosity towards the media, their leaders were quoted as stating "those who fight against us by the pen will die by the sword." The GIA also targeted foreigners in Algeria and killed at least 120 of them. Since the French supported the Algerian Government, France was also targeted with attacks taking place in Paris in 1995.[131]

In the 2008 Annual Threat Assessment of the Intellegence Community for the Senate Armed Services Committee, AQIM is called a "robust affiliate" of Al Qaeda and, as the most active terrorist group in northwest Africa, it poses a significant threat to U.S. and European interests in the area. There are also indications from Al Qaeda's central command in Southwest Asia that the

Libyan Islamic Fighting Group has united with Al Qaeda and accepted their leadership. The use of improvised explosive devices (IEDs) has been a staple tool of these terrorist groups. However, AQIM's use of advanced IEDs suggests that information on weapons technology is being imported from Iraq.[132] Following the Madrid bombing on 11 March 2004, and the formal transformation of AQIM, some analysts believe that this insurgency must be evaluated in the context of international Islamist militancy. Emily Hunt stated that the "expanding operations throughout Sahelian Africa, funneling African volunteers to fight in the Iraq insurgency, connections in Europe, and increasingly sophisticated internet propaganda strategy are of particular concern."[133]

... AQIM's use of advanced IEDs suggests that information on weapons technology is being imported from Iraq.

While many of the attacks take place in urban areas, the operational area of AQIM is vast, desolate, and extremely hard to control. Much of the population of the southern Sahel is nomadic, constantly moving in search of water and meager grazing land for their herds. Unless one has traveled in these remote areas, it is difficult to conceptualize the immensity and barrenness that prevails. Imagine crossing areas in which there are no way stations or major landmarks for several hundred miles.[134] Given that complex environment, the potential for the establishment of new training areas for AQIM should be of concern to all of the countries of the region. Counterinsurgency operations in the Maghreb would require continuous vigilance, including airborne and space-based sensors, to monitor the remote areas. Such advanced technical capabilities are beyond reach of local governments and would necessitate Western support. Even their resources would be stretched.

The number of insurgents involved with AQIM is not known. Estimates run from several hundred to a few thousand fighters. The nature of the organization is such that there may well be splinter groups, with no formal affiliation with the central organization, that on occasion use the name. In addition, there are many people who are supportive of the counter-government political aims of the Islamic movements. It is known that the extreme and often indiscriminant violence of earlier insurgent organizations did disenfranchise many civilians. Despite their relatively small numbers, AQIM demonstrates the problems of irregular warfare in which an isolated organization has a major impact on the societies in which they operate and exert influence far beyond those borders.

In several countries, there are however, many young people who are attracted to the notion or ideology of jihad. Many areas of Morocco are seriously underdeveloped and many young people feel the government has not kept its promises of social and economic development.[135] Thus recruitment for jihad is easy and their focus is both anti-secular-government and anti-Western. This support for jihad is an expanding trend that is disturbing to local officials.[136] The combination of ideological appeal and social disenfranchisement offers AQIM a larger pool from which to recruit those who would transition from transient supporter to full-fledged terrorist.

Algeria has suffered the brunt of the AQIM attacks, as well as those from its predecessor organizations. In February 2007 they conducted seven coordinated bombing attacks in different cities within a 6-hour period. In April 2007 insurgents set off two car bombs in different locations, including a governmental palace at about the same time. These killed 33 people and wounded an additional 222. A third bomb was discovered and defused before it could detonate. That month the bombings were characterized as a "major escalation aimed at raising the group's profile."[137] On 9 September a suicide bomber in Batna, Algeria killed 19 people who were waiting to greet Algerian President Bouteflika and injured 97 others. There have been four such attempts on his life.

On 11 December two car bombs were detonated in Algiers. One struck the Algerian Supreme Court building. The other targeted the United Nations High Commissioner of Refugees killing 17 UN employees and 25 civilians, and wounding 158 others. These bombings continued into 2008, proving that the Algerian government was not capable of stopping them. In August an attack against police officers in Tizi Ouzou seemed specifically designed to demonstrate the limitations of the government.[138] Later in August 2008 multiple bombings targeting civilians resulted in 60 deaths. In these attacks, AQIM announced responsibility.[139]

The Kingdom of Morocco has long been considered one of the most moderate and peaceful states in North Africa. There have, however, been a number of attacks that have taken place in recent years. Moroccan officials are worried about the influence of jihadists that have been coming from the Gulf states. Long before the emergence of AQIM, they began moving into the area and intermingling with the local population. Some of them have married into Moroccan families and are raising families. There is concern that many of these foreign jihadists are in sleeper cells. Being more sophisticated than their Algerian counterparts, the Moroccan cells are using money from the Gulf countries to provide social services not available from the government to build a positive image. They are also working to enter into the political system and have obtained a small number of seats in parliament.

In May 2003 a major coordinated attack took place in Casablanca. Five suicide bombs were detonated within thirty minutes at different locations around Morocco's largest city. With the exception of a Jewish center, all of the targets were spots frequented by European foreigners. A total of 41 people were killed and over 100 injured in those bombings and ties between those attacks and Al Qaeda were suspected at the time.[140]

In the spring of 2007, three more bombings took place in Casablanca. On March 11th there was an event at an Internet café where the terrorists appeared to be searching for instructions on explosives. Some analysts believe this was an accidental detonation, and that they were preparing to attack another target at a later time. That notion was substantiated in early April when another counterterrorist operation ended with additional suicide detonations. One of those killed was a brother of the 11 March bombing suspect. Then, on 14 April another suicide attack took place, this one apparently targeting the U.S. consulate and cultural center. These events, and others like them, are indications that a threat is close to the surface in Morocco.[141]

Morocco is extremely vulnerable as any successful terrorist attacks could seriously jeopardize the thriving tourist industry that contributes greatly to the Moroccan economy. It is estimated that over 14 million foreign visitors, mostly European, came to that country last year and the government is actively pursuing even more. Those travelers contributed to about 19 percent of the country's GDP. Moroccan security efforts are considerable and largely effective.[142] However, a single attack against a tourist site, such as was initiated by Al-Gama'a al-Islamiyya at Luxor Egypt on 17 November 1997, would have devastating implications.

Worth noting is that the tactics of AQIM seem to be paralleling those used by Al Qaeda in Iraq during 2004 and 2005. That is, when they experience trouble striking hard targets such as U.S. troops, they began attacking soft targets that included extensive civilian casualties. At least four of their attacks in Algeria were aimed at the president of that country. Protection of key installations in Morocco has been effective. Thus, most of the casualties in AQIM attacks throughout the region have been innocent civilians. In Iraq, inflicting civilian casualties proved to be a strategic mistake as the local populace felt compelled to accept security measures from the U.S.[143] It is too soon to determine whether AQIM will continue with indiscriminate attacks in the Maghreb. It is clear that they will employ these measures in other areas of the world with no regard for innocent civilians.

There are other conflicts and terrorist affiliations in areas that complicate matters further. In addition to AQIM activities, there are reports that Hezbollah has some support in Northern Africa. That would bring other actors of the Middle East into play. While best known for their activities in Lebanon and Israel, Hezbollah has a global reach including the Western Hemisphere.[144] Conversely, fighters from the Maghreb have joined in battles in many other countries. In particular, Algerians have been involved in combat in Afghanistan, Lebanon, Iraq, Bosnia, Chechnya, and other locations around the world.[145]

Another long-running feud exists. For several decades much of Western Sahara has been occupied by Morocco. At odds are the Saharawi Liberation Front, known as the Polisario, and the Moroccan government, which sees the area as part of their territory. While a United Nations-brokered ceasefire is in effect, there are periodic outbreaks of hostilities between these factions. The Polisario wants full independence while Morocco adamantly opposes the idea. Resolution in the near-term is unlikely.

While not directly involved with the AQIM struggle, Western Sahara has been influenced by the U.S. aid provided to Morocco in that conflict. That assistance has bolstered Morocco's ability to maintain control of the territory. The issues are complicated and adversely impact the counterterrorism efforts against AQIM.

The European Connection

What happens in Africa doesn't always stay in Africa. Because of the prior colonial relationships, and current policies for political asylum, many Western European countries have experienced substantial immigration from Africa. Northern African countries have a geographical proximity that contributed extensively to the south to north migration. As might be expected, concomitant with these demographic shifts have come exportation of terrorism and the roots of the insurgencies emerging in the Sahel.

This is entirely consistent with the global objectives of Al Qaeda and their relationship to AQIM.

On 11 March 2004 terrorist struck the railroad system in Madrid Spain. The detonation of 10 IEDs on 4 trains resulted in the deaths of 191 people and wounding of more than 1,800 other commuters. This attack is on record as one of the major terrorist strikes this century. The perpetrators came from Morocco, Syria, and Algeria. The alleged mastermind, known as Mohammed the Egyptian, was arrested and later acquitted at trial. Extensive investigation revealed that while Al Qaeda was not directly involved in the bombing, they inspired it.[146] The concept of the superorganism has been previously discussed, but has direct bearing on this event. For an insurgent organization to be dangerous does not require a formal command and control system.

Al Qaeda's second in command, Ayman al-Zawahiri, has articulated the need for exportation of the insurgency from Northern Africa to Southern Europe. In a video released on 19 September 2007 he directed that a mission of AQIM was to restore *al-Andalus*, most of the Iberian Peninsula, to Muslim control. He specifically addressed conflict in Spain and France. Over the past few years several hundred suspected Islamic militants have been detained in France and Spain. Some of those have had ties to the organizations that were the forerunners of AQIM.[147]

It is known that AQIM is actively recruiting members in Europe to be trained for terrorist missions. Concerned U.S. and European officials note that they are looking for people who are second or third generation citizens and can easily blend in with people from the target countries.[148] Following the modus operandi of bin Laden in the 9/11 attacks, they are giving training to clean-shaven young men who dress in Western clothing. Arms smuggling from Europe to the Maghreb is also a problem. There is a report of an attempt to ship weapons worth €5million (Euros) from Belgium, Germany, and the Netherlands to Morocco.[149]

French authorities have reported a strengthening of the AQIM offshoot in France, Germany, and Italy. Spain has indicated active recruiting is occurring for training in North Africa along with collection of funds to support insurgent efforts. In August 2007 analysis by the Jamestown Foundation concluded that, "the globalization of the Algerian struggle merger between GSPC and Al Qaeda represents an increased threat for France in particular and Europe in general. AQIM has set up the means and motive: funding stream established throughout Europe; increased availability of (experienced) foot soldiers; and a renovated ideological framework able to inspire a new generation of fighters. The only element missing in the puzzle is opportunity."[150]

To address the AQIM insurgent threat in North Africa requires a broader perspective than the Maghreb terrain. The relationship between Al Qaeda and AQIM takes the earlier conflicts aimed at destabilizing the governments in North African countries to a new, more global level. Al Qaeda's objectives have always been to attack European and American interests wherever they can find them. Sympathetic immigrant populations in Europe, and a growing insurgency in North Africa, are a perfect combination to achieve their objectives.[151] It is also a situation that begs for American SOF intervention on many levels, but only with careful consideration of the desired outcomes and potential for unintended consequences.

Worth noting is that other areas of Africa are also exporting and importing their conflicts from abroad, and the U.S. is not immune from the problem. Somali diaspora in the United States and Canada appear to continue to support the discord. For example, Minneapolis, Minnesota has experienced a number of murders that may be related to old grievances.[152] Similarly Edmonton, Alberta, Canada has experienced an exceptionally high homicide rate among young Somali males who have immigrated there.[153] There are also concerns about the number of Somali immigrants who seem to be missing from their prior locations in North America. There are indications that some may have been recruited to become suicide bombers, returning to their home country.[154] While most violence is concentrated in specific geographic areas, IW clearly transcends all borders. Countering these threats requires synchronization of SOF elements of all combatant commands. This effort will extend beyond the Department of Defense to other national agencies and even state and local law enforcement cooperation.

China: The Other Elephant in the Room

Although this monograph is devoted to irregular warfare in Africa, there are many external participants that directly, or indirectly, influence those conflicts. It is impossible to ignore the elephant in the room—China. In recent decades China has slowly interjected itself into African affairs. As a technologically developing country, China recognized that it eventually would need to import both mineral and energy resources (oil).

In fact, this strategic maneuver has been executed adroitly. Initially, China provided African countries of interest with foreign aid and asked little in return. In Kenya, I was informed that the Africans preferred to get assistance from China rather than the United States. The quote regarding acceptance of assistance was, "They don't tell us what to do." That comment referred to the caveats that often come with American aid that attempts to dictate sexual practices and acceptable birth control techniques (anti-abortion rules).

The scope of China's intervention in Africa has become the focus of numerous scholarly articles, plus frequent items in the press. At the University of Southern California conference on public diplomacy in Africa, it was noted that contrary to some rumors, the U.S. Africa Command was not specifically created to counter growing Chinese involvement there. However, the growth of the Chinese relations with African countries has attracted considerable attention at the U.S. State Department. It even evoked comments from then-President Bush who noted that the U.S. does "not see a 'zero-sum' competition with China for influence in Africa."[155] Carrying that thought a step further, Deputy Secretary of State Negroponte noted "our goal is to engage China to try and define and expand a common agenda for Africa that ultimately will serve both our national interests…"[155]

While the State Department addresses a common agenda, other U.S. policy makers are becoming increasingly concerned about China's projection of *soft power* and the effect it has on America's economic and strategic interests. They note that the *deep pockets* of Chinese industry has advantages in that it has no shareholders and can afford short-term losses for strategic gain.[156] When this is viewed in light of the support for irregular warfare in Sudan, the practices of unrestricted loans and provision of weapons take on both tactical and strategic significance.

The State Department recognizes that China does not place restrictions on their foreign aid but raises a concern about that practice. In testimony before the Senate Foreign Relations Committee, State Department officials "warned China's assistance efforts in Africa, which emphasize 'no strings' are not predicated on the same kinds of conditionality as other countries' aid program, could endanger progress in promoting good governance and market reform in Africa."[157] This has direct bearing on irregular warfare, as these practices allow or enhance less competent governance, which in turn is likely to foster instability and rebellions.

The migration from China to Africa has involved tens of thousands of people over the past decade. There are now an estimated 750,000 Chinese in various African countries. In addition, over 900 Chinese companies are doing business there.[158] They are engaged in many large engineering projects, including dam building, road construction, laying railroad lines, and oil pipelines. For these projects Chinese companies insist on bringing in their own labor force, most of which remains isolated from the local population.

Rapidly developing, China has had double-digit economic growth for the past decade. To keep momentum, that requires increased energy. Unfortunately, China's domestic oil production is declining and there is insufficient coal output to meet demands.[159] In 2006 China's net oil imports were 3.5 million barrels a day. That is expected to rise to 13.1 million barrels per day by 2030, and the Chinese are already laying the groundwork for that increase. China now receives one-third of its oil from Africa.[160] They have already built a 500-mile highway and railroad from Khartoum to the Red Sea. That runs alongside the pipeline that carries the oil directly to waiting Chinese tankers.[161] Five countries in Africa account for 85 percent of the oil exports to China. Of those, three, Nigeria, Congo, and Sudan, have serious instability problems.[162]

Across Africa, China is involved in development of the infrastructure. According to the Council on Foreign Relations, "Experts say the roads, bridges, and dams built by the Chinese firms are low cost, good quality, and completed in a fraction of the time such projects usually take in Africa." In addition, they contributed peacekeepers, sent doctors, and have cancelled $10 billion in bilateral debts. However, they are also willing to pay bribes and underbid local companies on projects.[163] In the long run, accepting the lowest bid is not always the best move for African countries as their indigenous workforce often pays the price.

In addition to peacekeepers, China is providing arms and military advisors to several of its African allies. The serious *weapons for oil* deals in Sudan have already been addressed. However, there are several other countries that receive military assistance including Ethiopia, Eritrea, Burundi, Tanzania, Nigeria, and Zimbabwe.[164] Given the instability in some of these countries, delivery of

In addition to peacekeepers, China is providing arms and military advisors to several of its African allies.

additional weapons should be of concern to all with interests in the area. That is especially true for Zimbabwe, considering the extreme volatility since the failed election in June 2008.

While trade with China is rapidly increasing, there are intra-continental issues that are slowing that process. The transportation systems are poor as are the ports. High tariffs are placed on manufactured goods entering each country.[165] That leads to very inefficient border crossings where it is not uncommon for truck drivers to wait days for passage. That is particularly problematic with perishable foods that spoil before delivery.[166] Not only are these practices wasteful and time consuming, they increase corruption that goes beyond bribery to let goods pass. It also enhances the criminal network that smuggles weapons to insurgent groups in several countries.

The Heritage Foundation bluntly notes that China provides ideological support that gives despots international legitimacy and influence in the United Nations. The Chinese efforts help blunt Western pressures for democratic principles, economic openness, and human rights. They state, "Beijing succors would-be junta leaders and illiberal rebels who want power and would roll back political reforms in immature democracies. These rebels seem to believe that if they want to overthrow a legitimate government, China will work to bolster their international legitimacy in the United Nations and other international fora."[167] It is important to comprehend this influence in order to understand the full scope of irregular warfare in Africa.

Additional Factors Influencing Irregular Warfare in Africa

Corruption

Corruption is epidemic in many African countries. Angola and Zimbabwe rank among the worst on the continent, with Zambia, Madagascar, Mozambique, and Malawi not far behind.[168] Angola held its first election in 18 years in September 2008. Not much is likely to change, however, as this country can be classified as a *kleptocracy*. Angola is rich in natural resources and is the newest member of OPEC. Despite an annual economic growth rate averaging over 15 percent, the United Nations Development Program ranks Angola as 162nd out of 177 countries in human development. It is China's biggest supplier of oil, and is the second largest diamond producer in Africa. Yet, 70 percent of the people live below the poverty line and only 1 percent of the population enjoys the wealth from the mineral resources. It is claimed that 70 percent of the country's oil revenue is unaccounted for (at least $4 billion) and about 25 percent of the government's income "simply disappears." [169]

Following independence in 1975, after nearly 500 years of Portuguese presence and rule, Angola became embroiled in a civil war that lasted until 2002. The casualty figures vary, but at least half a million died, and several million were displaced.[170] Elections were held in 1992, but that devolved into continuation of the civil war that ended when Jonas Savimbi, the leader of National Union for the Total Independence of Angola (UNITA), was killed in combat. It should come as no surprise that the combatants in the civil war broke down along tribal and ethnic lines.

Of current concern is the political future of the exclave of Cabinda, sometimes known as "the Kuwait of Africa." This tiny disconnected northern province contains an estimated 65 percent of the oil in Angola. However, under the current central government, only 10 percent of the revenues from that oil stay in the province, which is geographically closer to Congo than it is to the rest of Angola. The possibility of a Nigerian-style oil insurgency has been raised as a means for gaining independence. Separatist insurgent forces have contested the area ever since 1975, and even earlier.[171]

It is noted that some analysts do not like the use of the word *kleptocracy*, but it is included as it accurately describes a concentration of wealth in the hands of the governing party. This acquisition is sustained only because of the government's ability to set policies and control all aspects of the economy. The definition of corruption, however, is a much broader term and may exist at any level of government or within the civilian business sector. Bribery, for instance, can occur within the confines of a legitimate government that is neither aware of the practice nor approves of it.

While not condoning corruption per se, Dr. Margaret Polski of George Mason University, and a former State Department official with extensive experience in Africa, noted that "Corruption is relatively harmless when an economy is consistently growing, increasing numbers of people are benefiting from activity in the domestic economy, and the perception that property rights are expanding to include more people over time." To support this claim she points to U.S. political and economic development over the past three centuries during which there was significant corruption and yet we eventually grew into a thriving democracy. The point was made to counter arguments that "Africa is a basket case," when significant progress has been made in many countries.[172]

However, generally the experience of corruption in Africa has been concentration of wealth in the hands of a very few, with nepotism and tribalism as key factors. When corruption is widespread and affects the daily lives of ordinary citizens, such as bribing bureaucrats for doing their job, or law enforcement officials to prevent arrests, there is a commensurate degradation in trust for all government institutions. Taken to extremes, as does occur, organized crime, with the complicity of police and courts, can make normal life nearly impossible. The long-term negative impact on government legitimacy is extraordinary.[173]

Unfortunately, the disparity between the wealthy and the poor is increasing and likely to exacerbate conflict in the future. From a SOF perspective, acceptance of corruption by foreign officials can lead to potential mission-related problems. U.S. laws tend to be intolerant of any corrupt practices and provide stiff penalties for any American, including both government employees and business persons, who attempt to circumvent them. While that may be idealistically attractive to academicians and politicians, it is often not realistic in foreign cultures that have centuries of history in which bribery was not only tolerated as a means of accomplishing tasks; it was and often is mandatory.

Drugs and Terrorism

Drugs and terrorism are inextricably related. Around the world, drugs are the number one funding source that supports terrorism. Africa is no exception to the influence that extensive drug money can bring. Due to increased enforcement pressure by the U.S. in the Caribbean in recent years, there have been substantial changes in the drug routes from South America to Europe. Major shipments flow from the Andean region, south through Paraguay, and into the Atlantic Ocean via ports in Argentina and Uruguay.[174]

These new routes bring the bulk shipments of narcotics across the South Atlantic to be transshipped from small countries in West Africa such as Sierra Leone, Liberia, Ivory Coast, Ghana, and Guinea-Bissau. As most of these countries are just coming out of extended periods of civil strife, narcotics smuggling, as a destabilizing factor, is extremely dangerous to fledgling governments. It is estimated that one quarter of the illegal drugs headed for Europe pass through the area. While exact figures are not known, one educated guess by the UN Office for Coordination on Humanitarian Affairs suggested 40 tons of cocaine was coming through Africa each year. The narcotics smuggling in West Africa is believed to account for 99 percent of the illicit drugs on the continent. Of further concern is that organized crime smuggling routes, once established, move other commodities including weapons and human trafficking.[175]

Observers believe that Guinea-Bissau, a failed state, has been taken over by drug lords and is the *first narco-state*. While ranking as the fifth poorest country in the world, the flamboyance of Colombian drug barons there readily stands out, and the value of their drugs exceeds the entire national income. The drug lords simply buy the necessary government services. The country's airspace is unpatrolled, and it does not have enough money to provide the police with radios, phones, or even gas for their vehicles. The drug lords come equipped with high-tech tools, and they have even developed unmanned submarines that can be programmed to make the trans-Atlantic runs with no risk to their personnel.

Reenergizing the ancient slave-trade routes seems logical. It is the shortest route from South America to Europe. Once in Guinea-Bissau, the drugs move through Morocco on the old cannabis trail, or sometimes via air. One flight to Amsterdam in 2006 was found to be carrying 32 human mules. An indication of what makes this so lucrative is that the UN Office on Drugs and Crime now estimates that the UK and Spain have overtaken the U.S. in cocaine consumption on a per head basis.[176]

The bad political situation deteriorated even further in March 2009. On Sunday, the 1st of the month, the army chief of staff, General Batista Tagme Na Waie was killed by a powerful bomb blast at his headquarters in Bissau. Less than a day later military personnel attacked and killed President Joao Bernardo "Nino" Vieira as he tried to flee his home.[177]

It is because of such violence that there has been little Western diplomatic presence in Guinea-Bissau. In 1998, as violence increased, both the U.S. and the United Kingdom shut their doors and moved their governmental activities to Senegal. In July 2007, in response to the growing narcotics trade, the U.S. finally restored a diplomatic office.

Also believed lurking in Côte d'Ivoire are militias that were never disarmed after the peace process of 2002–2003 between the rebel-held north and government-run south. The peace deal actually was not signed until

2007 and calls for elections in which both the government and former rebels can participate. While it is believed that security has improved, the farmers in west Côte d'Ivoire, the world's largest cocoa producers, are concerned about a reemergence of the murders and violence that took place a short time ago. The northern farmers, known as Burinabes, indicate they just want to get on with their agriculture.[178]

Cracks are already appearing in this recent truce. Part of the agreement includes providing national identity cards that become part of the voter registration process. Unfortunately there are significant differences between the sides as to how many people will be counted. The proximity of illegal drugs, guns, and money, coupled with questioned election procedures is worth watching.

While there has been a dramatic increase in drug smuggling in Western Africa, it is not the only region on the continent affected. As with other more affluent places in the world, South Africa has also experienced an increase in drug use as orchestrated by organized crime.[179] The country is both a user of drugs and a transit point for cocaine from South America and heroin from the Far East. Mandrex (methaqualone), which provides a quick euphoric state when smoked, is reportedly the preferred drug of abuse, and South Aficans are the fourth largest producer of cannabis in the world.[180]

Human Trafficking/Slavery

Slavery is making a comeback. A global issue with special concern for Africa is the rise of human trafficking to the third most lucrative enterprise for organized crime. The average price for each slave is now about $12,500. It is estimated that there are between 600,000 and 800,000 people trafficked each year, resulting in a profit to slave runners of a low-estimated $10 billion.[181] The Council of Europe states the profit is much higher at $42.5 billion, and noted it is easier to smuggle people than drugs.[182] Importantly, there is a correlation between irregular warfare and human trafficking that bears examination.

According to U.S. State Department definitions, there are several forms of trafficking in persons. These categories include child soldiers, forced labor, bonded labor, debt bondage and involuntary servitude among migrant laborers, involuntary domestic servitude, forced child labor, and sex trafficking and prostitution. They include two other child sex related crimes as well.[183] Examples of several of these categories have been mentioned in this paper, including child soldiers and forced labor. Also alleged has been sexual slavery, accompanied by some of the most egregious rapes and crimes against women imaginable. The U.S. State Department uses a three-tier system to evaluate whether or not countries are attempting to eliminate human trafficking. Tier 1 implies the countries fully comply with the *minimum* standards. Only one African country, Madagascar, is in that category. The vast majority of African countries are in Tier 2, meaning they appear to be making an effort to improve. However, 17 of those countries are questionable and on a *watch list*. Two African countries, Algeria and Sudan, are listed in Tier 3 meaning they are not even attempting to improve. Finally, there is a *Special Watch List*, which contains five African countries.[184]

Human trafficking is well documented in Western Africa. Less is known about the problem in the East and Horn of Africa, but there is significant trade. The biggest issue regarding the slaves was lack of governmental acknowledgement. The regional contributing factors include extreme poverty, lack of female education, poor data, and armed conflict.[185]

Use of children as soldiers is continuing and is not restricted to insurgents. In several cases, governments are involved in recruiting youngsters to fill the ranks. Chad was named in September 2005 as violating a UN Security Council admonishment pressuring them to stop using child soldiers.[186] The Coalition to Stop the Use of Child Soldiers recently reported that children were engaged in armed conflict in 19 countries. The African countries listed were Burundi, Central African Republic, Chad, Côte d'Ivoire, the Democratic Republic of Congo, Somalia, Sudan, and Uganda. In addition to combat, these children are also used as spies, informants, and messengers. The group also noted that in Kenya and Nigeria criminal organizations use children for political purposes.[187]

A child soldier of the DR Congo. Photo courtesy Lt Col Rudy Atallah, U.S. Air Force.

There is no doubt that human trafficking is expanding globally and African countries contribute greatly to the problem. From an irregular warfare perspective kidnapping children has proven to be a cheap and efficient means of recruiting foot soldiers who are considered expendable to the insurgent leadership. Of course the psychological effects are devastating, but that is of no consequence to their commanders. Similarly, thousands of people have been forced to work as laborers to provide logistical support to the insurgents. The alternative to refusal to comply is death. Women are pressed into service and used as sex slaves for troops that stay in remote areas. For some groups, the slave trade is a profitable method to raise funds to support their movement.

Democratization/Elections

Recent years have seen a number of elections around the continent. The results have been far from perfect, and some offer reason for concern. The violent response to close elections in Kenya has already been discussed and offers one example of the potential volatility these events can bring.

Similarly, the elections in Zimbabwe evoked another crisis. There the incumbent simply refused to accept the vote and refused to leave office. Holding the loyalty of the military, he was able to offer the winning opposition a weakened position in a power-sharing arrangement. These negotiations are likely to drag on for an extended period, possibly years. While violent confrontations have been minimized, Zimbabwe is in an economic disaster from which it will take a long time to recover.

Neighboring Zambia also held elections in which the acting incumbent president, Rupiah Banda won by about a 2 percent plurality. These elections were held due to the unexpected death from a stroke of President Levi Mwanawasa, and the rival, Michael Sata, was widely expected to win. Despite allegations of vote rigging, and while the opposition contested the election in court, they did not resort to a violent confrontation as they had previously.[188] Considering the experiences in Zimbabwe and Kenya, this election can be taken as a positive sign in the move toward true democratic processes.

In late November 2008 elections were held in Nigeria. In the city of Jos in the center of the country, rumors emerged that Christians had won all of the seats in a predominantly Muslim section of the city. This led to rioting resulting in about 400 fatalities from intra-communal violence.[189]

For the purposes of this paper, the important issue is not just the casualties in this city that has been the scene of repeated religious conflicts this past decade. The point is that more people died in Jos, Nigeria than were killed during the exact same time as the terrorist attacks in Mumbai, India, and yet the incident went almost unreported by the international press. While it is certainly true that the terrorist attacks in Mumbai had unique characteristics worthy of significant attention, even there the Western media placed great emphasis on the relatively few European and American fatalities. There is little doubt that the incidents in Jos were very ugly. However, expectations of election violence, and strong media bias against concentrated reporting such incidents in Africa, suggest that most of the world will not be made aware of election-related violence unless the situation devolves into a conflagration.

For SOF operators, there should be an awareness of increased cultural sensitivities in both pre and post election periods. There is increased danger to personnel assigned on missions in those countries. As indicated in these examples, the concept of peaceful transition of power from one group to another is not fully integrated. This also allows for the potential for SOF members to become unexpectedly embroiled in local or regional politics without any intention of so doing.

Challenges and Opportunities

Many factors complicate involvement in Africa. For one, corruption is epidemic. While this can be opposed, corruption at all levels of governments will remain a problem for the foreseeable future. The issue of tribalism will not subside in the near term. As noted, in several countries, this is and will remain the most fundamental factor that determines relationships, economics, and especially governance. Probably the best that can be hoped for is some improvement in basic human rights and equality of opportunity for all citizens.

Other interrelated factors that will impact the future of Africa include poverty, disease, and population. Many difficult issues are related to poverty. Several countries have population growth rates that are unsustainable based on resources available. As noted, despite wealth from natural resources, the general population of these countries rarely benefits. Rather, there is a concentration of wealth in a relatively small sector, and often much of that is foreign controlled. It is an unfortunate fact that while overall global wealth is increasing, so is poverty. Africa and South America offer extreme examples.

Except in selected metropolitan areas, the education systems are substandard when compared to Western schools. This means that there are not enough skilled workers to create a vibrant industrial infrastructure. Throughout the continent, it is not uncommon to find foreign senior managers operating a variety of industries in which the African laborers perform low-skill, manual tasks, and at minimal wages. There may also be tribal or ethnic disparity between managers and workers. Tribal affiliation may determine the amount and type of education that is available to an individual. When perceived as unjust, this situation can be a breeding ground for discontent and even insurgency.

In many areas of Africa, literacy is relatively low. That will impact the types of training materials that can be used. Written manuals, the standard for American forces, may be of little value. Training aids must be designed to meet the skill level of the trainees. That content extends beyond troops being trained to civilians in the operational area. Of course, experienced SOF personnel know that literacy does not necessarily correlate with intelligence. Indigenous people usually understand their environment exceedingly well as it is imperative to their survival. We can learn a lot from them if we just listen.

Part and parcel with poverty are diseases that have already been controlled in the developed world. Noted has been that millions of children die every year from preventable diseases. Of special consideration is the AIDS epidemic that has devastated several countries. While this does decrease the population, it has also eliminated much of the experienced workforce in the affected countries. From a SOF perspective, operating in these conditions is challenging and entails considerable risks.

Because of the historic delineations, border disputes will continue to plague African international relations and impact the foreign policies of external nations. When tribalism is interjected into the equation, no simple resolutions are readily available. This presents difficult considerations for determination of what aid should be provided and to whom.

There are no simple explanations that can be used to describe the problems found in Africa. This diverse and complex continent will offer extreme challenges in foreign policy for all future U.S. administrations. There is little doubt that our national interests are involved in many areas of the continent. However, every major power on earth also has recognized the importance of Africa and the resources it can provide. For the coming decades, complex international juxtapositions will be a norm, and the stakes very high.

Here, possibly more than in other areas of the world, the chess metaphor for engagement of competitors is inappropriate. In fact, chess, with its moves and countermoves, would be relatively simple compared to the realities at hand. More useful might be the ancient Asian game of *Go*, in which opponents seek to surround and isolate each other's positions. To win, one must have strategic awareness of the entire situation yet respond tactically to current situations. Part of the strategic difficulty of the game stems from finding a balance between such conflicting interests. Players strive to serve both defensive and offensive purposes and choose between tactical urgency and strategic plans. Thus it is in Africa. There is a need to directly confront and combat terrorists and insurgents, yet simultaneously develop the long-term relationships that will insure continued access to resources and support of mutually beneficial objectives.

Specifically, the interests and involvement of China in African affairs infers a strategic move designed to provide them access to energy including oil, natural gas, and uranium. In addition, many other natural raw materials will be aggressively sought after as global industrialization continues to grow. The U.S., Europe, and India will all be competing for those same resources.

Consideration must be given to the entire constituency of groups deemed as adversaries. As indicated in the section on Darfur, despite disclaimers, there is strong evidence that Russian pilots were flying bombing missions for Sudan. Whether these were active-duty air force or contract employees is in question. While the U.S. does not currently have armed forces deployed in the Darfur region, more active involvement is not implausible. A situation such as that could bring U.S. forces into direct confrontation with Russians, or other nationalities with whom we are currently in a state of peace (or at least not engaged in war). The massive proliferation of contractors may well complicate future operations and all possibilities must be taken into account.

To operate effectively across this continent will require a highly skilled, extremely flexible force. Operations, especially those designed to increase internal stability, cannot be done in an ad hoc basis by units on reprieve from combat duties in the Middle East. It will take a truly professional and dedicated effort if real success is contemplated.

The environment in which they will operate is as complex as any on earth. In many areas of Africa, SOF elements will encounter social, ethical, and even legal challenges to their operations. They must be able to work within constraints that are fundamentally different from their own social structures and values. Nationally and for senior leadership there will be tough choices regarding which values take priority.

… SOF elements will encounter social, ethical, and even legal challenges to their operations.

As examples of these tough choices, When working within tribally-based cultures, which groups do you train and support? Is it possible, through the development of military professionalism, to instill in our partners the concepts of equality among all people? What level of corruption in foreign establishments can be tolerated in order to accomplish specific missions, such as elimination of terrorist elements? Do we support certain countries, or groups, when border disputes arise? What are acceptable environmental risks when access to minerals is desirable? Do we contribute to maintaining poverty in large groups for the sake of gaining those materials? Will support to current military leaders be used to suppress political opposition?

The simple answer is to accept no tolerance. However, our foreign competitors in Africa have demonstrated that they will play by different rules and values. If the strict moral code is maintained, do we accept that the result may be to concede territory or resources to countries that are willing to overlook even major indiscretions? Conversely, if lower ethical standards are applied, the leadership we are attempting to raise to the higher standards will see us as hypocritical. What emerges is a moral dilemma with no easy answers.

Implications for SOF

At the beginning of this monograph *irregular warfare* was defined "as a violent struggle among state and non-state actors for legitimacy and influence over relevant populations." The significance of IW has moved it to a front-burner strategic issue within the Department of Defense, which resulted in a directive putting these efforts on par with conventional warfare and direction to USSOCOM to develop SOF capabilities for extending U.S. reach, operating with and through indigenous foreign forces.[190] In addressing strategic SOF roles, Colonel Joe Celeski, U.S. Army (Retired), noted efforts for preventing conflict by supporting regional engagement and insuring stability and security when possible.[191] As indicated in the following paragraphs, Africa offers significant challenges and opportunities for strategic SOF interventions.

By now the reader should know that conflict in Africa represents the epitome of IW. With the notable exception of the World War II traditional battles between the German and Allied forces, IW has been embodied in every hostility that has occurred there in the past century. Nearly every scenario in which SOF forces might engage in conflict on that continent will be IW. Even the training missions are designed to counter IW and enable African nations to establish and maintain security.

Much of the African continent is in turmoil, and is likely to remain so for the foreseeable future. With the creation of Africa Command, the United States unified our efforts and went on record as upgrading our national interests in the area. The mission statement is: "United States Africa Command, in concert with other U.S. Government agencies and international partners, conducts sustained security engagement through military-to-military programs, military-sponsored activities, and other military operations as directed to promote a stable and secure African environment in support of U.S. foreign policy."[192]

As connoted in the mission statement, the efforts in Africa will be interagency in nature, more so than any other combatant command. The issues elucidated in this report suggest that many of the military operations in Africa will be SOF oriented. That will be especially true for the development and strengthening of stability of existing military forces. However, recent evaluations of counterterrorism efforts have not been glowing. A Government Accounting Office (GAO) report on the *Trans-Sahara Counterterrorism Partnership* notes that the State Department-led, multiagency program suffered because three agencies developed separate plans related to their perspective partnership activities. While the plans show some collaboration, "they do not provide an integrated, comprehensive approach to guide the program overall."[193]

There are many challenges facing SOF operators on missions to African countries. From prior colonial rule there are two dominant languages,

English and French. In the areas previously occupied by the French and Belgians, the most prominent official language is French, often to the relative exclusion of English-speaking natives. In addition, there are hundreds of indigenous languages and dialects, and some people speak them exclusively. While some effort is made to have Special Forces linguistically qualified, they are far short of the numbers necessary to be conversant with host nation soldiers. Adding to the problems of providing written training materials, in some areas literacy of enlisted soldiers and civilians is very low.

While SOF operators may not have much choice in who they train, there are strategic issues that senior leadership will need to address at several levels. Stated U.S. national interests speak to supporting national governments if democratically elected. In reality, determination of legitimacy of governments may not be as simple as it sounds. There have been elections, such as during 2008 in Zimbabwe in which the outcome was simply stolen by incumbent Robert Mugabe. Other elections, such as during 2007 in Kenya were harder to determine and did devolve into the massive riots previously described. It should be noted that one of our allies in North Africa, Morocco, is still a kingdom even though it does have a functioning parliament.

As an example of the problems that can arise with military cooperation, in August 2008 a military coup in Mauritania, led by General Mohamed Ould Abdel Aziz, replaced President Sidi Ould Cheikh Abdallahi, who was elected just the year prior. It was the firing of several senior generals that prompted the coup. In recent years the Trans-Sahara Counterterrorism Partnership provided training to troops in that country.[194]

The pervasiveness of tribalism has been noted repeatedly in this paper. Since ethnic or tribal relationships often determine employment, it is inevitable that majority groups are usually going to dominate the military ranks, and especially the senior officer corps. Therefore, by training those soldiers, it may well be that we are ensuring the capability of the majority ethnic group to more handily exert control over subordinate or minority groups that live in their country. Professionalism is not as easily taught as is marksmanship or fire and maneuver drills. After all, it has only been a few decades since U.S. military units were racially integrated, and that was not without substantial pain and effort.

At the heart of this issue is the ultimate loyalty of the military. Many countries in Africa have histories of military coups. Some of those leaders, such as Mohmmar Qadaffi and Robert Mugabe, have been successful in retaining power for a long period of time. American Special Forces trained some coup leaders, including Sergeant Samuel Doe, who led the 1980 overthrow in Liberia. A decade later he was also deposed and brutally executed. It is difficult to determine the loyalties and intent of the troops provided for training. The only thing that can be done is to choose as wisely as possible, remembering that not all troops who seem friendly, and are willing to be trained, share common values with our soldiers.

Unfortunately, many SOF Mobile Training Teams are deployed on rather short duration missions, thus yielding limited exposure to the cultural environment. Given the complexity of the history and sociology of most indigenous areas of the world, it is impossible for team members to understand the environment in which they must conduct training or to develop the desired interpersonal relationships that are borne out of mutual trust.

This is far from a new problem, and has resurfaced in nearly every conflict in which the U.S. has participated. While SOF elements often pride themselves on regional knowledge, it would take more than a lifetime to fully understand the intricacies and ambagiousness of Africa's indigenous societies. The best that can be done is to comprehend the basic social constructs and their histories, and to be aware that you don't understand everything there is to know about the situation. It is the nature of Americans to want to make friends. Other cultures view these situations differently. A quote worth remembering is that *countries do not have friends; they have interests.* That sensitivity will minimize the probability of becoming involved in adverse situations that could harm our national interests.

The ongoing war in Iraq and Afghanistan further exacerbates the problem of in-depth understanding for foreign societies. The omnipresent operational tempo of those operations has seriously degraded the foreign area capability for every other region of the world. Attempts are made at retaining socially sensitive skills related to other areas, but life and death exigencies of combat appropriately take top priority. Cultural awareness and interpersonal relationships are perishable and retained only through repetitive engagement. The appropriate balance of assignments represents tough choices.

An emerging SOF element that will play increasingly important roles in Africa is the Marine Special Operations Command (MARSOC). Historically the U.S. Marines have been called upon to execute many missions in

Africa. Some of the best known have been the evacuation of American and European civilians when hostilities threatened all foreigners. In particular, the Marine Special Operations Advisory Groups (MSOAGs) are suited for foreign internal development missions, much like their U.S. Army Special Forces counterparts. Also expanding are the Marine Special Operations Battalions that will be involved in direct action and special reconnaissance missions in Africa.[195]

Many of the physical environments in which SOF must operate in Africa are extraordinarily harsh and inhospitable. Troops operating in Sahara or the Kalahari often find themselves facing extreme heat, lack of access to water, and occasional gale-force winds with blowing sand that can penetrate most any piece of equipment and make life miserable. Much of Central Africa provides a totally different, but equally difficult environment. In this tropical region the dense forests and extensive rainfall can greatly inhibit cross-country travel and extract a physical toll on all non-indigenous personnel attempting to operate there. The people residing in these remote areas and with whom SOF units may work often live on the margins of civilization. Therefore, long-duration assignments, especially if repetition is anticipated, are generally infeasible.

Disease is another significant concern. The extreme environmental conditions of Central Africa have spawned some of the deadliest diseases known to man including Ebola, Hemorrhagic Fever, and Marburg virus. Scientists believe that some of these rare but often fatal illnesses first evolved in monkeys and then migrated to humans through close contact between the species. While controversial, AIDS may have been passed to humans by the same means. Whatever the origin, the AIDS pandemic kills about two million people each year in Africa. Other common diseases that are nearly unheard of in the developed world include sleeping sickness, yellow fever, cholera, river blindness, and tick bite fever.[196] Polio, once thought eliminated, is making a comeback in Africa and a resurgence of malaria accounts for about one million deaths, many of which are children under five years of age.[197] West Africa also spawns illnesses rarely seen in other areas of the world. Among them is Lassa disease, an arenavirus which is generally borne by rodents and transmitted to humans who come in close proximity.

Based on the presence of these virulent diseases, careful medical planning, preparation, and monitoring for all SOF operators deploying to these areas is imperative. That monitoring must continue well after they return home or transition to another theater. Some of the exotic neurodegenerative African diseases can be introduced into the soldier's body and remain dormant for extended periods of time before onset of symptoms.

In addition, these diseases also impact the soldiers being trained in affected areas. One of the common problems with troops in underdeveloped areas of the world is marksmanship. Poor diet often leads to impaired eyesight. Rarely do they have access to competent optometry services and do not get the glasses that would easily correct the vision problems.

It is acknowledged that there is a direct linkage between terrorism and organized crime. In both Southern Africa and Western Africa this has become a significant problem as some areas have become transit points for drugs shipped from South American en route to Europe. Weapons smuggling is also a serious problem. One issue that has emerged between the U.S. and our African allies is differentiation between criminal and terrorist activities. It is argued that some governments label all adverse activities as terrorism with the anticipation that the U.S. will provide funds to them from our counterterrorism initiatives.

Unlike NATO allies or other countries that have been supplied with U.S. weapons, many African countries have acquired materiel from other sources. Therefore equipment compatibility can be a challenge. SOF trainers must know the foreign systems and be able to understand how to integrate disparate equipment and make it all functional. Under such circumstances, maintenance and logistics are particularly difficult.

Finally, there is the issue of race. As a fully integrated military, the United States assigns people based on skill and merit. Race is rarely considered as a qualification. Except for restricted skill sets, even gender is not a major factor in determination of assignments. However, when superimposed on societies that take these issues very seriously, the introduction of a person with the *wrong* background could affect acceptance of the unit's advice. There are, of course, tradeoffs that may be made based on how urgently the recipient believes that they need the resources provided by the U.S. SOF element. Examples abound indicating that these issues can be overcome in many cases. For example, American units in predominantly Muslim countries have assigned women as advisors to men, a custom not easily accommodated. For isolated SOF elements, it may be efficacious to take race, gender, and even religion into account when making sensitive assignments. If access and acceptance is key to success, consideration of these factors may be necessary whether or not they comply with established DoD policies. For much of the world, mix and match does not apply to personal relationships.

There is little doubt that SOF elements will play key roles in support of African military development. They are by far the best suited for missions ranging from foreign internal development, to stability operations, and when necessary small scale direct action operations.

Summary

Comprised of 53 separate and disparate countries, Africa offers complex challenges to U.S. foreign policies, especially those related to defense. Clearly, there are national interests involving security and availability of energy resources and in countering terrorism that manifests global latitude. Only a few short decades out of colonial rule, many of the countries suffer from weak governance. Corruption, a global problem, is epidemic in many of these fledgling democracies and threatens social evolution into a more stable form of management. The concentration of wealth, derived by a governing few who exploit the country's natural resources, has been labeled kleptocracy.

While reduced in numbers, there are several ongoing conflicts based on historical grievances exacerbated by artificially created states with boundary delineation designed long ago for the convenience and self-interests of extra-continental powers. With more than 400 indigenous tribes, plus genetic infusion from Europe and Asia, the population of the continent evinces extreme heterogeneity.[198] As indicated, fierce intertribal rivalries have led to horrendous casualties. Disconcerting is that even evidence of genocide can be ignored both intra-continentally, and by culturally astute people in the more technically developed Western world.

Prior to the opening of the Suez Canal, Africa was strategically located on European naval routes for passage to Asia, thus attracting people to the port cities. In addition, Africa is rich in natural resources, proven to be both a blessing and a curse. While the lure of gold and diamonds once brought Europeans in their search for wealth, oil has become the principal commodity that attracts foreign interest and investment. The continuously increasing demand for energy has produced global competition, and the U.S. has specifically designated security of some African petroleum as a national interest. Oil has been located in several areas of the continent, but that is usually followed by regional instability.

With 22 percent of the world's landmass, Africa has other large mineral deposits including uranium, chromium, manganese, bauxite, cobalt, platinum, copper, and gold. In addition, the hardwood forests of the Congo basin provide lumber for a voracious global market, but bring worry about depletion of this critical, oxygen producing resource. In parts of Africa reside unique, sometimes endangered, species of animals that require protection in order that the lucrative, burgeoning tourism industry can flourish.

Problematic, however, is that despite the abundance of these resources, poverty among the common people is growing, and Africa holds several countries with the poorest people on the planet. In addition to government corruption, population growth, environmental changes (including desertification), rampant debilitating (and often fatal) diseases, inadequate communication and transportation systems, and limited educational opportunities contribute to the impoverishment. Of course, all of those factors become amplified when juxtaposed with widespread armed conflicts, which in recent years have displaced tens of millions of people from their homes, and directly contributed to the deaths of millions more. Certainly these conditions can easily lead to disenfranchisement and a propensity to join or support an insurgency.

While Europeans have historically viewed Africa as their exclusive domain, China has emerged as a global economic player with concerted interests there. As seen throughout this paper, the influence of China is felt in many areas and ranges from providing *no strings* economic assistance and development of infrastructure to circumventing arms embargos while supporting despots via its actions in UN Security Council. While the United States has participated in many projects, recent policy objectives indicate that involvement will increase and that our relations with China vis-à-vis Africa will be one of mutual cooperation. Their actions in Sudan put that notion to a severe test.

Across the continent irregular warfare is a norm, and there are several conflicts worth attention. The crisis in Darfur and Chad still rages with no end in sight. American assistance has focused on providing substantial food supplies for refugees, while EU and AU allies provide minimal peacekeeping forces that are at risk in the area. On the horizon is a renewed civil war between northern and southern Sudan, probably precipitated by the mandatory elections regarding independence of the south. Some experts believe that the situation could become worse than Darfur. The conflict in Congo remains unresolved and tens of thousands of people die each month from both direct and indirect results of that civil war. The situation in the Horn of Africa is still volatile and will not subside soon. In the Niger River Delta, MEND threatens oil production while in northern Niger the Tuaregs impact the uranium mining operations. Across the Maghreb AQIM continues to recruit, train and cause problems for governments in the region. It also exports terror to Europe and other destinations around the world.

While there are two current wars generating genocidal levels of casualties, Darfur and Congo, it is highly probable a third will reignite within the next two years. The UN-brokered peace treaty for the civil war between northern and southern Sudan called for multi-party elections and a referendum on separation. That civil war provides another nightmare scenario combining historic indigenous racial, ethnic, religious, and tribal issues juxtaposed with external competition for energy resources and infusion of modern weapons. Nearly forgotten is that the T-72 tanks on the hijacked ship, while initially headed for Mombasa, were really destined for delivery in South Sudan in preparation for the next round of warfare.

The new illegal-drug routes from South America are a destabilizing factor in West Africa as the narcotics money overwhelms the local economies. The increased drug smuggling is also adversely impacting South Africa. Coups and failed elections plague both Mauritania and Zimbabwe. Angola held open elections that resulted in the incumbent being returned to power even though the country is a kleptocracy. The outcome was never really in doubt as all of the media was controlled by the government. Of concern is what happens in the oil-rich exclave of Cabinda, and what role Angola may choose to play in supporting the current government of Congo as that fight drags on. The ethnic-based strife that followed Kenya's 2008 election is a reminder of just how quickly relative stability can devolve. It demonstrates that tribal tension, always close to the surface, can turn deadly almost instantaneously. Across all of these conflicts, gender related atrocities play a significant role in perpetuating violence.

While many SOF training missions have been conducted in African countries, most were for relatively short durations. American response to repeated social and environmental catastrophes has been hindered by lack of adequate bases from which to operate or provide large-scale logistical support. Given the significant concerns by many African governments about the ultimate intent of U.S. Government policies, that situation is unlikely to change. Therefore, many missions requiring immediate intervention will probably be relegated to MARSOC units operating from, or supported by, the U.S. Navy.

However, it will be more important for SOF to move beyond crisis intervention and rescue missions. Gaining and maintaining confidence is an arduous process that demands constant attention. High on the AFRICOM priority list of operations should be increased and sustained foreign internal development training missions with militaries that are considered relatively stable. Whenever possible repeated face to face contact should be initiated to both instill the values of a professional forces and to provide for early indications of impending danger and conduits for mitigating outbreaks of violence.

Another critical instrument for gaining trust and confidence will continue to be civil support missions such as medical civil action programs (MEDCAPs) and engineering projects. With burgeoning desertification providing a well is invaluable and goes a long way in demonstrating the utility of American presence and support of our values. On a continent with 11 percent of the world's people, and 1 percent of the medical care, SOF medical elements are worth their weight in gold.

The Horn of Africa provides a special set of problems with global implications, and the potential for SOF intervention. Specifically, the recent increase in high-value ship hijackings has caused the shipping industry to reconsider transiting the Suez Canal and avoiding the Gulf of Aden in favor of the far longer voyage around the treacherous waters of the Cape of South Africa. While not exclusively an American problem, the U.S. Navy SEALS are specifically trained for ship boarding and counterpiracy missions.

Given the history of American interventions, and sometimes lack thereof (Rwanda), in Africa, the Middle East, and elsewhere, development of better relations between the U.S. and most African countries will be painstakingly slow. It remains unlikely that Africa Command will be able to secure a permanent headquarters anywhere on the continent, and will be consigned to frequent trips to meet with leaders of the countries. However, SOF elements have already established relationships in many countries and offer the best hope for expanded contacts, especially as junior leaders from African militaries mature and move into increasingly responsible positions. Building on the personal relationships our SOF personnel create through repeated deployments can have strategic value in obtaining the goals and objectives of the United States in general and Africa Command in particular.

If the U.S. is going to increase its engagement in Africa, then competent strategic thought is required. Assisting in capacity building and support for emerging democracies, is at best tricky. Given the level of operational commitment in Iraq and Afghanistan, there are few combat troops left to respond to other crises. The guidance of the AFRICOM objective "to keep American combat troops off the continent for fifty years" leaves limited options. There seems to be little doubt that Africa provides a full spectrum SOF operational environment. Great care and planning is required, but SOF offer the best toolkit for the job.

Endnotes

1. While the formal definition of *irregular warfare* is relatively new, the concepts are not. However, the asymmetric nature of recent battles has brought new light onto the topic as U.S. military forces evolved their countermeasures, sometimes by trial and error. In September 2007 the Department of Defense published the *Irregular Warfare Joint Operating Concept*. It defined *irregular warfare* (IW) "as a violent struggle among state and non-state actors for legitimacy and influence over relevant populations." They note that IW is inherently protracted and favors the indirect and asymmetric approaches. It also acknowledges that the challenges will require application "of all available instruments of national power." The concept notes that IW is about people, not about which side has the biggest guns, or most advanced technology. Success depends on understanding social dynamics, tribal politics, religious influences and cultural mores. In December 2008, a new DoD directive further codified the importance of IW.
2. Senior joint commander, addressing the SOC Africa Command regional meeting held at MacDill Air Force Base, 23 May 2008.
3. Former principal Deputy Undersecretary of Defense for Policy, conference entitled *AFRICOM: The American Military and Public Diplomacy in Africa*, University of Southern California, 8 February 2008.
4. An ambassador assigned to AFRICOM, conference entitled *AFRICOM: The American Military and Public Diplomacy in Africa*, University of Southern California, 8 February 2008.
5. Mike Bottoms, "USSOCOM celebrates its 20th Anniversary," USSOCOM Public Affairs Office press release, May, 2007, http:/socom.mil/Release/May/20thAnniv/USSOCOM%celebrates%20its20%2020thAnniversary.pdf.
6. U.S. Forces Command, *Joint Operational Environment (JOE 2008),* http://jfcom.mil/about/facts_joe2008.html.
7. John B. Alexander, Ph.D., "Intelligence Scotomas in Central and South America," *The Proteus Management Series*, Vol 1, Issue 4, March 2008.
8. Thomas L. Friedman, *Hot, Flat and Crowded: Why We Need A Green Revolution - And How It Can Renew America*, (Farrar, Straus and Giroux, 2008), p.155.
9. Christopher Clapham, "The Challenges of Peacekeeping in Africa." *The African Military in the 21st Century, Report of the 2007 Tswalu Dialogue*, The Royal United Service Institute, May 2007
10. Chris Lowe, et. al, "talking about 'Tribe' - Moving from Stereotypes to Analysis," *Africa Action*, February 2008, http://www.africaaction.org/bp/documents/TalkingaboutTribeFeb2008Update_001.pdf.
11. There will undoubtedly be controversy concerning the use of the word *tribe*. I have chosen to include it as a topic as my experience in Africa, and that of many others, finds that tribal identity is extremely important in many areas on the continent. It is used by many Africans to describe themselves. As everyone is aware, it is a term that is frequently used in European and American media. The

social revisionists who argue against the use of the word claim that it promotes racial distortions and by fiat portrays "ethnicities as primitive and savage." They rightfully point out that other factors, such economics and politics, are integrally juxtaposed in current conflicts. However, they ignore that many of the economic and governance issues, in fact brake down along tribal affiliations. It is anachronistic to regard all tribes as "primitive." Despite the controversy, it must be noted that even, "African leaders see tribalism as a major problem in their countries." For reference see note 12.

12. "Rwanda: How the genocide happened," *BBC News*, updated 1 April 2004, http://news.bbc.co.uk/2/hi/africa/1288230.stm.
13. John B. Alexander, discussions and observations made during personal travel in Rwanda in November 2007.
14. Gerard Prunier, "Rwanda's ghosts refuse to be buried," *BBC News*, 6 April 2009, http://news.bbc.co.uk/go/fr/-/2/hi/africa/7981964.stm.
15. Jeffrey Gettleman, "Disputed Vote Plunges Kenya into Bloodshed," *The New York Times*, 31 Dec 2007.
16. "U.N.: 600,000 Displaced in Kenya Unrest," *CBS News*, 11 Feb 2008, http/:www.cbsnews.com/stories/2008/02/11/world/main3815702.shtml.
17. "Kenya diplomatic push for peace," *BBC News*, 2 January 2008, http:/news.bbc.co.uk/2/hi/Africa/7167363.stm.
18. U.S. State Department, Bureau of Consular Affairs, "Travel Warning-Kenya," 21 March 2008, http:/www.travel.state.gov/travel/cis_pa_tw/tw/tw_923.html.
19. "Kenya: Mungiki Claims 1,000,000 Members," *Nairobi Star*, 17 December, 2008, http://allafrica.com/stories/2008/12160756.html.
20. "FACTBOX_ Key facts about Kenya's Mungiki gang," *Reuters*, 5 June 2007, http://www.alertnet.org/thenews/newsdesk/L05868431.htm.
21. "Profile: Kenya's secretive Mungiki sect," *BBC News*, 24 May 2007, http://news.bbc.co.uk/go/pr/fr/-/2/africa/6685393.stm.
22. "More killings feared as Kibaki vows new Mungiki war," *Nairobi Chronicle*, 7 December 2008, http://nairobichronicle.wordpress.com/2008/12/07/%ef%bb%bfmore-killings-feared-as-kibaki-vows-new-mungiki-war/.
23. "Morgan Tsvangirai," Biography of, Answers.com, www.answers.com/topic/morgan-tsvangirai.
24. ""Mugabe orders white farmers to leave," *BBC News*, 12 August 2002, http://news.bbc.co.uk/2/hi/africa/2187453.stm.
25. Martin Ngwenya, "Keep Reinforcing Zimbabwean Capacities to End Cholera Epidemic - WHO," *Voice of America*, 31 March 2009, http://www.voanews.com/english/Africa/Zimbabwe/2009-03-31-voa59.cfm.
26. Peter Brookes and Ji Hye Shin, "China's Influence in Africa: Implications for the United States," *The Heritage Foundation, Backgrounder*, 22 February 2006 http://www.heritage.org/research/asiaandthepacific/bg1916.cfm.
27. "Zimbabwe inflation rockets higher," *BBC News*, 19 August 2008, http://news.bbc.co.uk/go/pr/fr/-/hi/business/7569894.stm.

28. "South Africa protests anti-immigrant violence," *Reuters*, 24 May 2008 http://www.msnbc.msn.com/id/24804749/.

29. The basis for this paragraph came from the comments provided by an unidentified, but obviously knowledgeable, double-blind review of the original manuscript at JSOU. I have modified the comments slightly.

30. Greg Campbell, "'Blood Diamond' shows how industry leaders turned a blind eye to the atrocities funded by their policies." *The Christian Science Monitor*, 11 Dec 2006. http://www.csmonitor.com/2006/1211/p09s01-coop.htm.

31. John B. Alexander, *Winning The War: Advanced Strategies, and Concepts for the Post-9/11 World*, (New York: St. Martin's Press, 2003).

32. Jon Swain, "Africa, China's new frontier," *The Sunday Times*, 10 February 2008, http://www.timesonline.co.uk/tol/news/world/africa/article3319909.ece.

33. "Africa Fast Running Down Resources – WWF Report," *World Environmental News*, 10 June 2008, http:/www.planetark.com/avantgo/dailynewsstory.cfm?newsid=48713.

34. Central Intelligence Agency, *The World Factbook*, updated 10 June 2008, https://www.cia.gov/library/publications/the-world-factbook/geos/eg.html.

35. Ibid.

36. John B. Alexander, personal observation in Rwanda, November 2007.

37. Dith Pran, "The Cambodian Killing Fields," http://www.dithpran.org/killing-fields.htm.

38. "Genocide in the 20th Century," *The History Place*, http://www.historyplace.com/worldhistory/genocide/bosnia.htm.

39. U.S. Department of State, Bureau of African Affairs, *Background Note: Sudan*, July 2008, http:///www.state.gov/r/pa/ei/bgn/5424.htm.

40. Brian Steidle, *The Devil Came on Horseback*, (New York: PublicAffairs, March 2007).

41. "Somalia: No Mercy in Mogadishu," *Africa Watch, Physicians for Human Rights*, 26 March 1992 http://www.hrw.org/reports/1992/somalia/.

42. Pauline Jelinek, "U.S. Special Forces in Somalia," *The Associated Press*, 10 January 2007, http://www.washingtonpost.com/wp-dyn/content/article/2007/01/10/AR2007011000438.

43. Alex Perry, "U.S. Kills Bin Laden's Man in Somalia, *Time*, 1 May 2008, http://www.time.com/time/world/article/0,8599,1736736,00.htm.

44. Shashank Bengali, "Islamist rise in Somalia is latest worry for U.S.'" *Miami Herald*, 23 November 2008, http://www.miamiherald.com/news/world/story/783915.html.

45. David Montero, "War-torn Somalia braces for fresh violence amid shaky peace accord," *The Christian Science Monitor*, 20 August 2008, http://www.csmonitor.com/2008/0819/p99s01-duts.html.

46. Jason McLure, "Somalia Islamist Says Ethiopia, Oil Exploration Deepening Crisis," *Bloomberg*, 27 August 2008, http://www.garoweonline.com/artman2/publish/Features_34.
47. "Pirates push for ransom bonanza," *Independent Online*, http://www.int.iol.co.za/general/newsprint.php?art_id=nw20081125175054912C360530.
48. Rob Crilly, Pirates help fund Somali warlords," *The Christian Science Monitor*, 27 August 2008, http://www.csmonitor.com/2008/0827/p06s01-woaf.html.
49. Brianna Blake, "Kenya Puts $150 Million Estimate on Payments to Somalia's Pirates," *Voice of America*, 21 November 2008, http://www.voanews.com/specialenglish/2008-11-21-voa1.
50. "Somali pirates seize more vessels," *BBC News*, 6 April 2009, http://news.bbc.co.uk/go/pr/fr/-/2/hi/africa/7985374.stm.
51. Alisha Ryu, "Pirates Reaching Farther Out to Sea to Hijack Vessels," 7 April 2009, *Voice of America*, http://www.voanews.com/english/2009-04-07-voa32.cfm.
52. Brian Steidle, In private conversation with the author in Las Vegas, 9 June 2007, Steidle, a former USMC officer, served as a UN observer in Sudan. He reported that they were aware that Russian pilots were flying the bombing missions and targeting the remote villages.
53. "Russian pilots fly Sudan MIGs in Darfur missions," *World Tribune*, 29 May 2008, http://www.worldtribune.com/worldtribune/WTARC/2008/af_suda0189_05_29.asp.
54. Hilary Andersson, "China 'is fueling war in Darfur'" *BBC News*, 13 July 2008, http://news.bbc.co.uk/go/pr/fr/-/hi/africa/7503428.stm.
55. Kristina Nwazota, "The Darfur Crisis," *PBS, The Online NewsHour*, 16 July 2008, http://www.pbs.org/newshour/indepthcoverage/africa/darfur/origins.html.
56. "ICC Prosecutor presents case against Sudanese President, Hassan Ahmad Al Bashir, for genocide. Crimes against humanity and war crimes in Darfur," *International Criminal Court*, The Hague, 14 July 2008, http://www.icc-cpi.wt/press/pressreleases/406.html.
57. "Warrant issued for Sudan's leader, " *BBC News*, 4 March 2009, http://news.bbc.co.uk/go/pr/fr/-/2/hi/africa/7923102.stm.
58. "Sudan: NGO Expulsions Fact Sheet #5 (FY2009)," *U.S. Agency for International Development*, 2 April 2009, http://www.reliefweb.int/rw/rwb.nsf/db900SID/MUMA-7QV33J?OpenDocument.
59. "Arab leaders snub a-Bashir warrant," *Al Jazeera*, 31 March 2009, http://english.aljazeera.net/news/middleeast/2009/03/2009330175846714662.html.
60. Nick Robertson, "Rape as a way of life for Darfur's women," *CNN*, 19June 2008, http://edition,cnn.com/2008/WORLD/africa/06/19/darfur.rape.
61. Abra Pollock, "Sudan: Civilians Driven Into Camps, Then The Bush," *Inter Press Service*, 16 November 2007.

62. Stephanie McCrummen, "A Wide-Open Battle For Power in Darfur," *Washington Post*, 4 July 2008, http://www.washingtonpost.com/wp-dyn/content/article/2008/07/03/AR2008070303915.html.
63. "Nightmare in Africa: the crisis in Darfur has spread to Chad," *Sunday Herald*, 2 February 2008, http://sundayherald.com/misc/print.php?artid=2014806.
64. Lydia Polgreen, "Chad Shuts Down Capital as 1,000 Rebels March In," *The New York Times*, 3 February 2008, http://www.nytimes.com/2008/02/03/world/africa/03chad.html.
65. Stephanie McCrummen, "Peacekeepers in Darfur Hobbled by Need," *Washington Post*, 20 June 2008, http://www.washingtonpost.com/wp-dyn/content/article/2008/06/19/AR2008061903552.html.
66. Mons Stava, Major in Norwegian Army, private conversation with the author about his experience while assigned as a UN staff officer on the peacekeeping mission in Darfur in 2007, Paris, France, 16 June 2008.
67. *Sunday Herald*, 2 February 2008.
68. "Influential power against Darfur Peace" *Sunday Nation*, 31August 2008, http://www.nation/co/ke/News/africa/-/1066/465024/-/149gf19z/-/index.html.
69. Moushumi Das Chaudhury, "UAE's total investment in Sudan hit Dhs25b," *The Bor Globe Network*, 30 August 2008, http://www.borglobe.com/business/business/12-business/957-uaes-total-investment-in-sudan-hits-dhs25b.
70. United Nations Office for the Coordination of Humanitarian Affairs, "Southern Sudan: A Path To 'Indivisible Peace' In The Country," in *Ten Stories the World Should Hear More About*, 2007, http://www.un.org/events/tenstories/07/sudan/shtml.
71. "Beyond The Latest North-South Crisis," *International Crisis Group*, Brussels, 13 March 2008, http://allafrica.com/stories/200803130746.htm.
72. Danna Harman, "How China's support of Sudan shields a regime called 'genocidal.'" *The Christian Science Monitor*, 26 June 2007, http://www.csmonitor.com/2007/0626/p01s08-woaf.html.
73. Andrew S. Natsios, "Beyond Darfur: Sudan's Slide Toward Civil War," *Foreign Affairs*, May/June 2008, http://www.foreignaffairs.org/20080501faessay87306/andrew-s-natsios/beyond-darfur.htm.
74. Ibid.
75. Ibid.
76. Michael Gordon and Jeffrey Gettleman, "U.S. Officials Say Israel Struck in Sudan," *The New York Times*, 27 March 2009, http://www.nytimes.com/2009/03/27/world/africa/27sudan.html.
77. "How Israel Foiled an Arms Convoy Bound for Hamas," *Time*, 30 March 2009, http://www.time.com/time/world/article/0,8599,1888352,00.html.
78. Alexis Okeowo, "Forgiving the Lord's Resistance Army," *Time*, 10 November 2007 http://www.time.com/time/world/article.0,8599,1682747,00.html.

79. David Smook, "Uganda/Lord's Resistance Army Peace Negotiations," *United States Peace Institute*, Feb 2008, http://www.usip.org/pubs/usipeace briefings/2008/0204 uganda.html.

80. "Who is re-supplying the LRA?" *BBC News*, 4 April 2009, http://news.bbc.co.uk/go/pr/fr/-/2/hi/africa/7983143.stm.

81. "Uganda rejects more rebel talks," *BBC News*, 19 August 2008, http://news.bbc.co.uk/go/pr/fr/-/hi/africa/7571139.stm.

82. Jeffery Gettleman, "Rebels Kill Nearly 200 in Congo, U.N. Says," *The New York Times*, 30 December 2008, http://www.nytimes.com/2008/12/30/world/africa/30ugganda.html?em.

83. "Congo Civil War," *Global Security*, http://www.globalsecurity.org/military/world/war/congo.html, accessed 25 August 2008.

84. "DR Congo uranium mine collapses, *BBC News*, 12 June 2004, http://www.bbc.co.uk/2/hi/africa/3887373.stm.

85. "IAEA probes DR Congo 'uranium smuggling," *ABC News* (Australia) 9 March 2007, http://www.abc.net.au/news/stories/2007/03/09/1867218.htm.

86. Michael Gavshon and Drew Magratten, "War Against Women," *60 Minutes, CBS News*, 13 January 2008, http://www.cbsnews.com/stories/2008/01/11/60minutes/3701249.shtml.

87. Craig Timberg, "Report: Congo's War and Aftermath Have Killed 5.4 Million," *Washington Post*, 23 January 2008, http://www.washingtonpost.com/wp-dyn/content/article/2008/01/22/AR2008012203340_pf.html.

88. Joe Bavier, "Congo war-driven crisis kills 45,000 q month-study," *Reuters*, 22 January 2008, http://www.alertnet.org/thenews/newsdesk/L22080212.htm.

89. "DR Congo pygmies 'exterminated,'" *BBC News*, 6 July 2004, http://news.bbc.co.uk/go/pr/fr/-/2/hi.africa/3869489.stm.

90. "Democratic Republic of Congo: Mass rape – time for remedies," *Amnesty International*, http://www.amnesty.org/en/library/asset/AFR62/081/2004/en.ba62821f-a3b8-11dc-9d08fi45a8145d2b/afr620182004en.pdf.

91. Robert Lloyd, "Rape in the Congo won't go unheard," *Los Angeles Times*, 8 April 2008, http://articles.latimes.com/2008/apr/08/entertainment/et-congo8.

92. Olivia Ward, "Sexual Violence 'a tactic of war.' UN Council says," *The Toronto Star*, 20 June 2008, http://www.commondreams.org/archive/2008/06/20/9759/.

93. Patrick Vinck, et al. "Living With Fear: A Population-Based Survey on Attitudes About Peace, Justice, and Social Reconstruction in Eastern Democratic Republic of Congo," *International Center for Transitional Justice*, August 2008 http://allafrica.com/peaceafrica/Resources/view/00011604.pdf.

94. Louis Charbonneau, "Peacekeepers may have committed sex abuse in Congo," *Reuters*, 13 August 2008, http://africa.reuters.com/top/news/usnBAN325952.html.

95. Michael Gavshon and Drew Magratten, "War Against Women," 13 January 2008.

96. Lydia Polgreen, "Fighting in Congo Rekindles Ethnic Hatreds," *New York Times*, 10 January 2002, http://www.nytimes.com/2008/01/10/world/africa/10congo.html.

97. "Rebels and army clash in DR Congo," *BBC News*, 28 August 2008, http://news.bbc.co.uk/go/pr/fr/-/2/hi/africa/7586617.stm.

98. "'Human Catastrophe' Grips Congo," *BBC News*, 31 October 2008, http://news.bbc.co.uk/go/pr/fr/-/1/hi/world/africa/7701269.stm.

99. "Reports: Angola promises troops as Congo peace fades," *CNN*.com, 12 November 2008, http://www.edition.cnn.com/2008/WORLD/africa/11.12/congo.fighting.

100. "Rwanda arrests Congo rebel leader," *BBC News*, 23 January 2009, http://news.bbc.co.uk/go/pr/fr/-/2/hi/africa/7846339.stm.

101. "Congolese flee widespread unrest," *BBC News*, 1 April 2009, http://news.bbc.co.uk.go/pr/fr/-/2/hi/africa/7977493.stm.

102. Brian Lionberger, Major, U.S. Army, "Emerging Requirements for U.S. Counterinsurgency: An Examination of the Insurgency in the Niger River Delta Region," Thesis for the *U.S. Army Command and General Staff College, Ft. Leavenworth, Kansas*, 2007 http://stinet.dtci.mil/cgi-bin/GetTRDoc?AD=ADA471380&Location=U2&doc=GetTRDoc.pdf.

103. Simon Robinson, "Nigeria's Deadly Days" *Time*, 14 May 2006, http://www.time.com/time/printout/0,8816,1193987,00.html.

104. "Biafra War," Global Security, http://www.globalsecurity.org/military/world/war/biafra.htm accessed December 2008.

105. Brian Lionberger, p. 49.

106. Tom O'Neill, "Curse of the Black Gold: Hope and Betrayal in the Niger River Delta" *National Geographic*, Feb 2007, http://ngm.nationalgeographic.com/ngm/0702/indea.html.

107. Stephanie Hanson, "MEND: The Niger River Delta's Umbrella Militant Group," *The Council on Foreign Relations*, 22 March 2007, http://www.cfr.org/publication12920/.

108. Sebastian Junger, "Blood Oil: Politics & Power," *Vanity Fair*, February 2007, http://www.vanityfair.com/politics/features/2007/02/junger200702.

109. "Nigeria's Shadowy Oil Rebels" *BBC News*, 20 April 2006, http://news.bbc.co.uk/1/hi/world/africa/4732210.stm.

110. Stephanie Hanson, 22 March 2007.

111. John Alexander, "The Evolution of Conflict Through 2020: Demands on Personnel, Machines, and Missions," *Global Trends 2020 Project_*, National_Intelligence Council, Alexandria, VA 25 May 2004. Superorganisms are comprised of many smaller parts but act in a manner that is greater than the sum of those parts. However, instead of complex command and control systems, each part responds to simple, often binary, instructions that facilitate complex actions. This concept runs counter to the traditional notion of centralized command and control and suggests a more complex target set when attempting to eliminate terrorist networks.

112. Brian Lionberger, op cit pp 78-80.
113. "Nigerian attack closes oilfield," *BBC News*, 20 June 2008, http://news.bbc.co.uk/2/hi/world/africa/7463288.stm.
114. Sebastian Junger, *Vanity Fair.*
115. Chindu Offor, "Nigeria's Niger Delta Kidnappings Total Nearly 130 Since 2008," *Voice of America*, 6 April 2009, http://www.voanews.com/english/Africa/2009-04-06-voa42.cfm.
116. "Nigeria Rebels Attack Pipelines." *BBC News*, 17 March 2009, http://news.bbc.co.uk/go/pr/fr/-/2/hi/africa/7947997.stm.
117. Hector Igbikiowubo, "Ex-Generals fingered as crude theft worsens: 680,000 barrels stolen per day," Vanguard, 27 March 2009, http://www.vanguardngr.com/context/view/32093/41/.
118. "Nigeria lost $20.7 bln in Niger delta unrest," *Agence France-Presse*, 10 April 2009, http://news.yahoo.com/s/afp/20090410/wl_africa_afp/nigeriaoilunrest/.
119. "Q&A: Tuareg Unrest," *BBC News*, http://news.bbc.co.uk/2/hi/africa/6982266.stm.
120. John Alexander, personal observations throughout southern Mali, including Timbuktu, in February and March 2006.
121. Bram Posthumus, "Niger: Along History. A Brief Conflict, an Open Future," *Searching for Peace in Africa, 2000* The European Centre for Conflict Prevention, http://www.conflict-prevention.net/page.php?id=40&formid=73&action=show&surveyid=37.
122. Kalifa Keita, Lt. Col. Army of Mali, "Conflict and Conflict Resolution in the Sahel: The Tuareg Insurgency in Mali," *U.S. Army Strategic Studies Institute*, 1 May 1998, http://www.strategicstudiesinsitute.army.mil/pdffiles/PUB200.pdf.
123. Tiemoko Diallo, "Tuareg Rebels in Mali Besiege Northern Garrison," *Reuters*, 14 September 2007, http://www.alertnews.org/thenews/newsdesk/L14864903.htm.
124. Tiemoko Diallo, "Gunmen attack Mali outpost, seize soldiers, weapons," *Reuters*, 19 July 2008, http://www.alertnews.org/thenews/newsdesk/L1981053.htm.
125. Tiemoko Diallo, "Suspected Tuareg rebels ambush Mali military convoy," *Reuters*, 28 August 2008, http://www.alertnews.org/thenews/newsdesk/L28877955.htm.
126. John Alexander, In February 2006, while in Timbuktu I met with the A-Team commander that was then training the Mali Army at that remote cite. The heavily dominant language of Mali is French. The commander indicated that there were two members of his team that could speak French, and only two members of the Mali unit that could speak English.
127. "A field officer of Niger's armed forces (FAN) Major Kindo Zada on Tuesday defected to the rebels," *African Press*, 22 July 2007, http://africanpress.wordpress.com/2007/07/22/a-field-officer-of-nigers-armed-forces-fan-major-kindo-zada-on-Tuesday-defected-to-the-rebels/.
128. "Niger's government, Tuareg rebels pledge peace, " *Reuters*, 7 April 2009, http://uk.reuters.com/article/homepagecrisis/id=UKL789895._CH_.2420.

129. Anneli Botha, *Terrorism in the Maghreb: The Transnationalisation of Domestic Terrorism*, Institute for Security Studies, June 2008.

130. Andrew Hansen and Lauren Vriens, "Al-Qaeda in the Islamic Maghreb (AQIM) of L'Organizsation Al-Qaïda au Maghreb Islamique (Formerly Salafist Group for Preaching and Combat or Groupe Salafiste pour la Prédiction et le Combat)" *Council on Foreign Relations*, 31 July 2008, http://www.cfr.org/publications/12717/.

131. Lauren Vriens, "Armed Islamic Group (Algeria, Islamists) (a.k.a. GIA, Groupe Islamique Armé, or al-Jama'ah al-Islamiyah al-Musallaha), *Council on Foreign Relations*, 22 June 2008, http://www.cfr.org/publications/9154/armed_group_algeria_islamists.html.

132. J. Michael McConnell, "Annual Threat Assessment of the Intelligence Community for the Senate Armed Services Committee," *Director of National Intelligence*, 27 February 2008, http:www.dni.gov/testimonies/20080227_testimony.pdf.

133. Emily Hunt, "Islamist Terrorism in North West Africa: A 'Thorn in the Neck' of the United States," *Policy Focus 65*, Washington Institute, February 2007, http://www.washingtoninsitute.org/templateC04.php?CID=266.

134. John Alexander, personal understanding from trekking in Mali, and motoring across the desert in Morocco convinced me that the notion of vastness can only be experienced, it cannot adequately be described.

135. Thomas Omestad, "The Casbah Connection: Why Morocco is producing some of world's most feared terrorists," *U.S. News and World Report*, 1 May 2005, http://www.usnews.com/usnews/news/articles/050509/9morocco.

136. John Alexander, private discussions with Commissioner Abdelali Zerrad, deputy chief of police, Rabat, Morocco concerning the evolution of AQIM and the current stability situation in that country, December 2007.

137. Eben Kaplan, "North Africa's Rising Radicals," *Daily Analysis, Council on Foreign Relations*, 16 April 2007, http://www.cfr.org/publications/13102/north_africa_rising_radicals.html.

138. Liam Stack, "Bombing Suggests Al Qaeda Remains a Threat in Algeria," *The Christian Science Monitor*, 6 August 2008, http://www.csmonitor.com/2008/0805/p99s04-duts.html.

139. "Al-Qaeda 'claims Algeria bombs,'" *BBC News*, 22 August 2008, http://news.bbc.co.uk/go/pr/fr/-/2/hi/africa/7576478.stm.

140. "Terror blasts rock Casablanca," *BBC News*, 17 May 2003, http://news.bbc.co.uk/go/pr/fr/-/hi/world/africa/3035803.stm.

141. "Morocco: The Unusual 'Bombing' in Casablanca," *STRATFOR*, 12 March 2007, http://www.stratfor.com/morocco_unusual_bombing_casablanca.

142. John Alexander, while traveling in Morocco in December 2007 and January 2008, the security measures were noticeable to a trained observer. Foreign groups were continuously shadowed by security personnel who always were advised of group itineraries.

143. "Algeria: AQIM and the Perils of Soft Targets," *STRATFOR*, 19 August 2008, http://www.stratfor.com/analysis/algeria_aqim_and_perils_soft_targets.

144. John Alexander, Proteus 2008.

145. John Alexander, *The Changing Nature of Warfare, the Factors Mediating Future Conflict, and Implications for SOF*, JSOU Monograph April,2006, p. 13. This international involvement in jihad is commensurate with the description of a philosophical model of war

146. "The accused in Spain's bombing trial," *MSNBC News Service*, 31 October 2007, http://www.msnbc.msn.com/id/2155905.

147. J. Peter Pham, "Al-Qaeda in the Islamic Maghreb: An Evolving Challenge in the War on Terror," *World Defense Review*, 8 May 2008, http://worlddefensereview.com/pham050808.shtml.

148. "New Al Qaeda Cell A Growing Threat to U.S." *CBS News*, 26 May 2008, http://www.cbsnews.com/2008/05/26/eveningnews/4127487.shtml.

149. "AQIM Tries To Set Up Operations in Morocco," *CBS News*, 7 August 2008, http"//www.cbsnews.com/blogs/2008/08/07/monitor/entry4327995.shtml.

150. Pascale Combelles Siegel, "AQIM Renews its Threats Against France," *Terrorism Focus*, Vol. 4, Issue 26, 7 August 2007, http://www.jamestown.org/terrorism/news/article.php?articleid=2373602.

151. Amir Taheri, "Al-Qaeda's sinister creep into North Africa" *The Times*, UK, 30 July 2008, http://timesonline.co.uk/tol/comment/columnists/guest_contributers/article4425820.ece.

152. Erin Carlyle, "Minneapolis Somali community facing a dark web of murders," *City Pages*, 12 November 2008, http://www.citypages.com/content/2008/11/12.news/minneaplois-somali-community-facing-dark-web-murders&page1.

153. Richard Liebrecht, "Somali community frustrated by deaths, language barriers," *Edmonton Sun*, 7 December 2008, http://edmontonsun.com/News/Alberta/2008/12/07/pf-7659381.html.

154. Sara A. Carter, "Somalis in the U.S. draw FBI attention," *Washington Times*, 29 December 2008, http://www.washingtontimes.com/news/2008/dec/29/somalis-in-us-draw-fbi-attention/.

155. Thomas J. Christensen and James Swan, Deputy Assistant Secretaries of State, "China in Africa: Implications for U.S. Policy," *Statement Before the Subcommittee on African Affairs of the Senate Foreign Relations Committee*, Washington, D.C. 5 June 2008, http://www.state.gov/p/eap/rls/2008/06/105556.htm.

156. Kerry Dumbaugh, "China's Foreign Policy: What Does It Mean for U.S. Global Interests?" Congressional Research Service, RL34588, 18 July 2008, http://www.fas.org/sgp/crs/row/RL34588.pdf.

157. Christensen, op cit.

158. Jon Swain, "Africa, China's new frontier," *The Sunday Times*, 10 February 2008, http://www.timesonline.co.uk/tol/news/world/africa/article3319909.ece.

159. Peter Brookes and Ji Hye Shin, "China's Influence in Africa: Implications for the United States," *The Heritage Foundation, Backgrounder*, 22 February 2006 http://www.heritage.org/research/asiaandthepacific/bg1916.cfm.

160. Stephanie Hanson, "China, Africa, and Oil," *Council on Foreign Relations, Backgrounder*, 6 June 2008, http://www.cfr.org/publications/9557/.

161. Swain, op cit.

162. Harry G. Broadman, "China and India Go to Africa," *Foreign Affairs*, Vol. 87 No.2, March/April 2008.

163. Hansen, op cit.

164. Ibid.

165. Broadman, op cit.

166. John Alexander, based on personal observation of multiple border crossing sites in Africa from 2006-2007.

167. Brookes, op cit.

168. "Zimbabwe: Corruption increasing, Transparency International," UN Office for the Coordination of Human Affairs, 8 October 2002, http://Irinnews.org/Report.aspx?Reportid=46595.

169. Peter Goodspeed, "Angola vote likely won't end corruption," *Financial Post*, 5 September 2008 http://financialpost.com/trading_desk/energy/story.html?id=770458.

170. "Angola's 25 Year Civil War" UN Security Council, http://www.globalpolicy.org/security/sanction/angola/001109.htm.

171. "Cabinda: Prospects for an Oil Insurgency in the Angolan Exclave," *The Oil Drum*, 13 May 2007, http://theoildrum.com/node/2536.

172. Margaret Polski, Ph.D., Private communication derived from review of an early draft of this monograph, 30 September 2008.

173. Also (see endnote 29) taken from the comments provided in a double-blind review of the original manuscript at JSOU.

174. Alexander, 2008.

175. "U.N.: Growing W. African drug trade may increase violence," *CNN*, 31 July 2008, http://edition.cnn.com/2008/WORLD/africa/07/31/africa.drugs.trade.ap/index.html.

176. Ed Vulliamy, "How a tiny West African country became the world's first narco state," *The Observer*, 9 March 2009, http://www.guardian.co.uk/world/2008/mar/09/drugstrade.

177. Gwynne Dyer, "Shoot-Out in Guinea-Bissau: Africa's 'narco-state' spins into violence as the result of its burgeoning drug trade," *Egypt Today*, April 2009, http://www.egypttoday.com/aspx?ArticleID=8465.

178. Ange Aboa Duekoue, "Security Better But Fear of Militias Lingers," *The Nation*, 17 August 2008, http://allafrica.com/stories/200808180105.html.

179. Anneli Botha, private conversations with this counterterrorism expert took place in Washington, D.C, 1 October 2008.

180. " International Narcotics Control Strategy Report," U.S. Department of State, March 2007, http/www.state.gov/p/inl/rls/nrcrpt/2007/vol1/html/80861.htm.

181. Ethan Kapstein, "Slavery Goes Global," *Policy Innovations, of the Carnegie Council,* November 2006, http://www.rolicyinnovations.org/ideas/commentary/data/slavery/.

182. "Council of Europe says human trafficking has reached 'epidemic proportions,'" *The Associated Press*, 5 December 2006, http://www.iht.com/articles/ap/2006/12/05/europe/EU_GEN_Greece_Trafficking.php.

183. "Major Forms of Trafficking in Persons," *U.S. State Department Trafficking in Persons Report*, 4 June 2008, http://www.state.gov/g/tip/rls/tiprpt/2008/105377.htm.

184. """Tier Placements," *U.S. State Department Trafficking in Persons Report*, 4 June 2008, http://www.state.gov/g/tip/rls/tiprpt/2008/105383.htm.

185. "Report of the Eastern and Horn of Africa Conference on Human Trafficking and Forced Labor," *African Network for Prevention and Protection against Child Abuse and Neglect*, Nairobi Safari Club, 5-7 July 2005, http://www.antislavery.org/homepage/resources/PDF/PDFtraffick.htm.

186. "Chad still using child soldiers – rights group," *The Irish Times*, 5 September 2008, http://www.irishtimes.com/newspaper/world/2008/0905/1220544891830.html.

187. "Child Soldier Global Report – Facts & Figures on Child Soldiers," *The Coalition to Stop the Use of Child Soldiers*, August 2008, http://www.childsoldierglobalreport.org/files_country_pdfs/Facts%20and%20Figures.pdf.

188. "Banda sworn in as Zambian leader," *BBC News*, 2 November 2008, http://news.bbc.co.uk/go/pr/fr/-/2/hi/africa/7704859.stm.

189. "Nigerian city tense after riots," *BBC News*, 30 November 2008, http://news.bbc.co.uk/go/pr/fr/-/2/hi/africa/7757398.stm.

190. Gordon England, Deputy Secretary of Defense, *Department of Defense Directive 3000.07,* 1 December 2008.

191. COL Joseph D. Celeski, U.S. Army (Ret.) "Strategic Employment of SOF in a UW Environment," *Special Warfare*, July-August 2008, p. 23.

192. GEN Kip Ward, Command Briefing, Partnership, Security, and Stability, United States Africa Command, 2008.

193. Katherine McIntyre Peters, "U.S. counterterrorism efforts flounder in North Africa, GAO says," *Government Executive*, 5 August 2008, http://www.govexec.com/dailyfed/08/08/080805kpl.htm.

194. "Troops stage coup in Mauritania," *BBC News*, 6 August 2008, http://news.bbc.co.uk/go/pr/fr/-/2/hi/africa/7544834.stm.

195. Matt Hilburn, "A Work in Progress: Marines continue to hone Special Operations role in SOCOM," *Seapower*, February 2008, http://seapower-digital.com/seapower/200802/pg=23.

196. Jason Beaubien, "AIDS, Other Diseases Create Poverty Trap in Africa," *National Public Radio*, 26 August 2006.

197. Thomas Nchinda, "Malaria: A Reemergence in Africa," *World Heath Organization, Geneva*, http://www.cdc.gov/ncidod/eid/vol4no3/nchinda.htm.
198. No one knows how many tribes there are in Africa, and some estimates place the figure over 1000. As indicated in prior notes, the definition of *tribe* is vague but the concept, as it applies to internal conflict, is a primary factor, and not a sign of intellectual laziness..

GPO U.S. GOVERNMENT PRINTING OFFICE: 2009–525–218